Discrete Signals and Systems with MATLAB®

Electrical Engineering Textbook Series

Discrete Signals and Systems with MATLAB®, Third Edition

Volumes in the series:

Volume I – Continuous Signals and Systems with MATLAB®
Taan S. ElAli

Volume II – Discrete Signals and Systems with MATLAB®
Taan S. ElAli

Discrete Signals and Systems with MATLAB®

Third Edition

Authored by

Taan S. ElAli

Embry-Riddle Aeronautical University

CRC Press
Taylor & Francis Group
Boca Raton London New York

CRC Press is an imprint of the
Taylor & Francis Group, an **informa** business

MATLAB® is a trademark of The MathWorks, Inc. and is used with permission. The MathWorks does not warrant the accuracy of the text or exercises in this book. This book's use or discussion of MATLAB® software or related products does not constitute endorsement or sponsorship by The MathWorks of a particular pedagogical approach or particular use of the MATLAB® software.

First edition published 2003
by CRC Press
6000 Broken Sound Parkway NW, Suite 300, Boca Raton, FL 33487-2742
and by CRC Press
2 Park Square, Milton Park, Abingdon, Oxon, OX14 4RN

First issued in paperback 2022

© 2021 Taan S. ElAli
CRC Press is an imprint of Taylor & Francis Group, an Informa business

No claim to original U.S. Government works

Visit the Taylor & Francis Web site at
http://www.taylorandfrancis.com

and the CRC Press Web site at
http://www.crcpress.com

ISBN: 978-0-367-54300-6 (pbk)
ISBN: 978-0-367-53993-1 (hbk)
ISBN: 978-1-003-08859-2 (ebk)

DOI: 10.1201/9781003088592

Typeset in Times LT Std
by Cenveo® Publisher Services

Visit the Routledge/eResources: www.routledge.com/9780367539931

Dedication

This book is dedicated first to the glory of Almighty God. It is dedicated next to my beloved parents, father Saeed and mother Shandokha. May Allah have mercy on their souls. It is dedicated then to my wife Salam; my beloved children, Nusayba, Ali and Zayd; my brothers, Mohammad and Khaled; and my sisters, Sabha, Khulda, Miriam and Fatma. I ask the Almighty to have mercy on us and to bring peace, harmony, and justice to all.

– Taan ElAli

Contents

Preface

All books on linear systems for undergraduates cover both the discrete and the continuous systems material together in one book. In addition, they also include topics in discrete and continuous filter design, and discrete and continuous state-space representations. However, with this magnitude of coverage, although students typically get a little of both continuous and discrete linear systems, they do not get enough of either. A minimal coverage of continuous linear systems material is acceptable provided there is ample coverage of discrete linear systems. On the other hand, minimal coverage of discrete linear systems does not suffice for either of these two areas. Under the best of circumstances, a student needs solid background in both of these subjects. No wonder these two areas are now being taught separately in so many institutions.

Discrete linear systems is a big area by itself and deserves a single book devoted to it. The objective of this book is to present all the required material that an undergraduate student will need to master this subject matter and to master the use of MATLAB® in solving problems in this subject.

This book is primarily intended for electrical and computer engineering students, and especially for the use of juniors or seniors in these undergraduate engineering disciplines. It can also be very useful to practicing engineers. It is detailed, broad, based on mathematical basic principles and focused, and it also contains many solved problems using analytical tools as well as MATLAB.

The book is ideal for a one-semester course in the area of discrete linear systems or digital signal processing, where the instructor can cover all chapters with ease. Numerous examples are presented within each chapter to illustrate each concept when and where it is presented. In addition, there are end-of-chapter examples that demonstrate the theory presented with applications using the MATLAB's Data Acquisition toolbox when applicable. Most of the worked-out examples are first solved analytically and then solved using MATLAB in a clear and understandable fashion.

To the Instructor: All chapters can be covered in one semester. The book concentrates on understanding the subject matter with an easy-to-follow mathematical development and many solved examples. It covers all traditional topics and a chapter on state space and FFT. If the instructor decides to cover block diagrams and the design of discrete systems, Chapters 7–10 are available *at* www.routledge.com/9780367539931 to cover those topics. At this link provided by the publisher, there are stand-alone chapters on sampling and transformations and two comprehensive chapters on IIR and FIR digital filter design in addition to the MATLAB m-files for chapters 1–6.

To the Student: Familiarity with calculus, differential equations and programming knowledge is desirable. In cases where other background material needs to be presented, that material directly precedes the topic under consideration (just-in-time approach). This unique approach will help the student stay focused on that particular topic. In this book, there are three forms of the numerical solutions presented using

MATLAB, which allows you to type any command at its prompt and then press the Enter key to get the results. This is one form. Another form is the MATLAB script, which is a set of MATLAB commands to be typed and saved in a file. You can run this file by typing its name at the MATLAB prompt and then pressing the Enter key. The third form is the MATLAB function form where it is created and run in the same way as the script file. The only difference is that the name of the MATLAB function file is specific and may not be renamed. Chapters 7–10 are not within this volume but are available online along with the MATLAB m-files for chapters 1–6 at www.routledge.com/9780367539931. These are stand-alone chapters on sampling and transformations and two comprehensive chapters on IIR and FIR digital filter design.

To the Practicing Engineer: The practicing engineer will find this book very useful. The topics of discrete systems and signal processing are of most importance to electrical and computer engineers. The book uses MATLAB, an invaluable tool for the practicing engineer, to solve most of the problems. Chapters 7–10 are not within this volume but are available online along with the MATLAB m-files for chapters 1–6 at www.routledge.com/9780367539931. These are stand-alone chapters on sampling and transformations and two comprehensive chapters on IIR and FIR digital filter design.

MATLAB® is a registered trademark of The Math Works, Inc. For product information, please contact:

The Math Works, Inc.
3 Apple Hill Drive
Natick, MA 01760-2098
Tel: 508-647-7000
Fax: 508-647-7001
E-mail: info@mathworks.com
Web: http://www.mathworks.com

About the Author

 **Taan S. ElAli,** PhD is a full professor of electrical engineering. He is currently the chair of the Engineering Program at ERAU. He has served as the chairman of the curriculum committee of the new College of Engineering at King Faisal University (KFU) from August 2008, when the committee was formed, to August 2009, when the final draft of the curriculum and its quality assurance plan for the college of engineering was approved by the University Council, the president of KFU, and the Ministry of Higher Education. From August 2009 to August 2010, Dr. ElAli was responsible for the implementation of the college of engineering curriculum quality assurance plan and the chairman of the electrical engineering department. Dr. ElAli has worked full-time at several academic institutions in the United States of America for more than 26 years in the areas of curriculum development, accreditation, teaching, research, etc. He received his BS degree in electrical engineering in 1987 from the Ohio State University, Columbus, Ohio, USA, his MS degree in systems engineering in 1989 from Wright State University, Dayton, Ohio, USA, and his M.S. in applied mathematical systems, and his PhD in electrical engineering, with a specialization in systems, controls, and signal processing from the University of Dayton in the years 1991 and 1993, respectively. Dr. ElAli has discovered a new approach to dynamic system identification and arrived at new mathematical formula in an automatically controlled dynamic system. Dr. ElAli is the author and coauthor of 13 books and textbooks in the area of electrical engineering and engineering education. He has contributed many papers and conference presentations in the area of dynamic systems and signal processing. In his efforts to revolutionize the engineering education, Dr. ElAli has contributed greatly to the Project-Based Curriculum (PBC) approach to engineering education. This approach has led to the restructuring of one of the major courses in the engineering curriculum and thus to the restructuring of the textbooks used for the course. From this restructuring, two textbooks have emerged: *Continuous Signals and Systems with MATLAB®* and *Discrete Systems and Digital Signal Processing with MATLAB®*. Universities like Rochester Institute of Technology, University of Texas at San Antonio, the University of Georgia, New Mexico State University, University of Massachusetts at Dartmouth, Texas A&M University at Kingsville, Western New England College, Lakehead University, University of New Haven, etc. have come on board to use these textbooks. He has taken the PBC approach to engineering education one step further and introduced student-focused learning through PBC. Dr. ElAli has contributed a chapter to the well-known *The Engineering Handbook* by CRC Press, another to the well-known *The Electrical Engineering Handbook*, by Taylor & Francis Group, and another two to *Advances in Systems*, and *Computing Sciences and Software Engineering*, by Springer, Germany, 2006

and 2007, respectively. Dr. ElAli is a Fulbright Scholar, and a senior member of the Institute of Electrical and Electronics Engineers (IEEE), a rare honor attained by fewer than 8% of IEEEs more than 400,000 members and is conferred only on those who have outstanding research achievements and who have performed great service to the scientific community.

Acknowledgment

I would like to thank the CRC Press team. Special thanks also go to Nora Konopka, who encouraged me greatly when I discussed this project with her for the first time, and has reaffirmed my belief that this book is very much needed.

1 Signal Representation

1.1 INTRODUCTION

We experience signals of various types almost on a continual basis in our daily life. The blowing of the wind is an example of a continuous wave. One can plot the strength of the wind wave as a function of time. We can plot the velocity of this same wave and the distance it travels as a function of time as well. When we speak, continuous signals are generated. These spoken word signals travel from one place to another so that another person can hear them. These are our familiar sound waves.

When a radar system detects a certain object in the sky, an electromagnetic signal is sent. This signal leaves the radar system and travels the distance in the air until it hits the target object, which then reflects back to the sending radar to be analyzed, where it is decided whether the target is present. We understand that this electromagnetic signal, whether it is the one being sent or the one being received by the radar, is attenuated (its strength reduced) as it travels away from the radar station. Thus, the attenuation of this electromagnetic signal can be plotted as a function of time. If you vertically attach a certain mass to a spring at one end while the other end is fixed and then pull the mass, oscillations are created such that the spring's length increases and decreases until finally the oscillations stop. The oscillations produced are a signal that also dies out with increasing time. This signal, for example, can represent the length of the spring as a function of time. Signals can also appear as electric waves. Examples are voltages and currents on long transmission lines. Voltage value gets reduced as the impressed voltage travels on transmission lines from one city to another. Therefore, we can represent these voltages as signals as well and plot them in terms of time. When we discharge or charge a capacitor, the rate of charging or discharging depends on the time factor (other factors also exist). Charging and discharging the capacitor can be represented thus as voltage across the capacitor terminal as a function of time. These are a few examples of continuous signals that exist in nature that can be modeled mathematically as signals that are functions of various parameters.

Signals can be continuous or discrete. We will consider only one-dimensional discrete signals in this book. A discrete signal is shown in Figure 1.1. Discrete signals are defined only at discrete instances of time. They can be samples of continuous signals, or they may exist naturally. A discrete signal that is a result of sampling a continuous signal is shown in Figure 1.2. An example of a signal that is inherently discrete is a set of any measurements that are taken physically at discrete instances of time.

In most system operations, we sample a continuous signal, quantize the sample values, and finally digitize the values, so a computer can operate on them (the computer works only on digital signals).

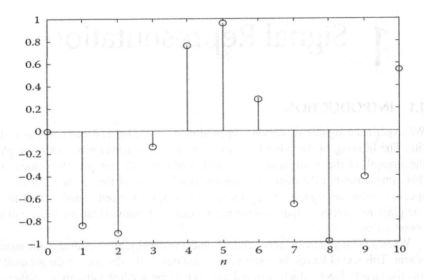

FIGURE 1.1   An example of a discrete signal.

In this book, we will work with discrete signals that are samples of continuous signals. In Figure 1.2, we can see that the continuous signal is defined at all times, while the discrete signal is defined at certain instances of time. The time between sample values is called the sampling period. We will label the time axis for the discrete signal as n, where the sampled values are represented at ... −1, 0, 1, 2, 3, ... and n is an integer.

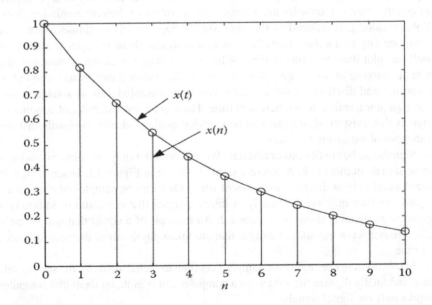

FIGURE 1.2   A sampled continuous signal.

1.2 WHY DO WE DISCRETIZE CONTINUOUS SYSTEMS?

Engineers used to build analogue systems to process a continuous signal. These systems are very expensive, they can wear out very fast as time passes, and they are inaccurate most of the time. This is due in part to thermal interferences. Also, any time modification of a certain design is desired, it may be necessary to replace whole parts of the overall system.

On the other hand, using discrete signals, which will then be quantized and digitized, to work as inputs to digital systems such as a computer renders the results more accurate and immune to such thermal interferences that are always present in analogue systems.

Some real-life systems are inherently unstable, and thus, we may design a controller to stabilize the unstable physical system. When we implement the designed controller as a digital system that has its inputs and outputs as digital signals, there is a need to sample the continuous inputs to this digital computer. Also, a digital controller can be changed simply by changing a program code.

1.3 PERIODIC AND NONPERIODIC DISCRETE SIGNALS

A discrete signal $x(n)$ is periodic if

$$x(n) = x(n + kN) \tag{1.1}$$

where
 k is an integer
 N is the period, which is an integer as well.

A periodic discrete signal is shown in Figure 1.3. This signal has a period of 3. This periodic signal repeats every $N = 3$ instances.

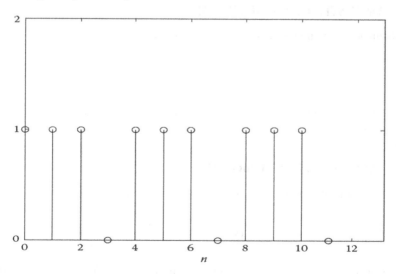

FIGURE 1.3 A periodic discrete signal.

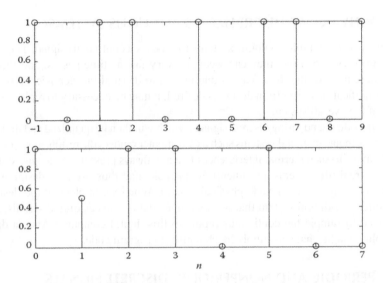

FIGURE 1.4 Signals for Example 1.1.

Example 1.1

Consider the two signals in Figure 1.4. Are they periodic?

SOLUTION

The first signal is periodic but the second is not. This can be seen by observing the signals in the figure.

1.4 UNIT STEP DISCRETE SIGNAL

Mathematically, a unit step discrete signal is written as

$$Au(n) = \begin{cases} A & n \geq 0 \\ 0 & n < 0 \end{cases} \tag{1.2}$$

where A is the amplitude of the unit step discrete signal. This signal is shown in Figure 1.5.

1.5 IMPULSE DISCRETE SIGNAL

Mathematically, the impulse discrete signal is written as

$$A\delta(n) = \begin{cases} A & n = 0 \\ 0 & n \neq 0 \end{cases} \tag{1.3}$$

where, again, A is the strength of the impulse discrete signal. This signal is shown in Figure 1.6 with $A = 1$.

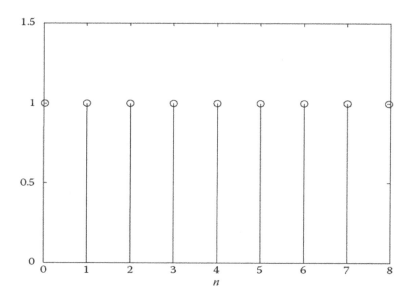

FIGURE 1.5 The step discrete signal.

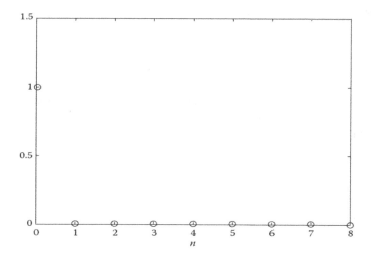

FIGURE 1.6 The impulse discrete signal.

1.6 RAMP DISCRETE SIGNAL

Mathematically, the ramp discrete signal is written as

$$Ar(n) = An \quad -\infty < n < +\infty \tag{1.4}$$

where A is the slope of the ramp discrete signal. This signal is shown in Figure 1.7.

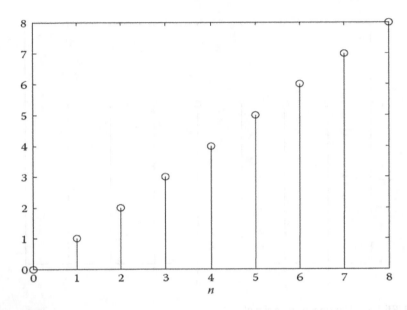

FIGURE 1.7   The ramp discrete signal.

1.7   REAL EXPONENTIAL DISCRETE SIGNAL

Mathematically, the real exponential discrete signal is written as

$$x(n) = A\alpha^n \tag{1.5}$$

when α is a real value. If $0 < |\alpha| < 1$, then the signal $x(n)$ will decay exponentially, as shown in Figure 1.8. If $1 < |\alpha| < \infty$, then the signal $x(n)$ will grow without bound, as shown in Figure 1.9.

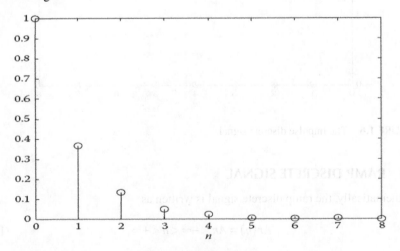

FIGURE 1.8   The decaying exponential discrete signal.

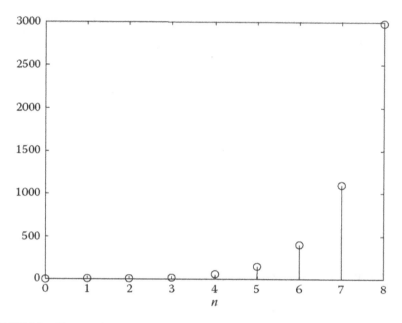

FIGURE 1.9 The growing exponential discrete signal.

1.8 SINUSOIDAL DISCRETE SIGNAL

Mathematically, the sinusoidal discrete signal is written as

$$x(n) = A\cos(\theta_0 n + \phi) \quad -\infty < n < +\infty \tag{1.6}$$

where
  A is the amplitude
  θ_0 is the angular frequency
  ϕ is the phase

A sinusoidal discrete signal is shown in Figure 1.10. The period of the sinusoidal discrete signal $x(n)$, if it is periodic, is N. This period can be found as in the following development.

$x(n)$ is the magnitude A times the real part of $e^{j\theta_0 n + \phi}$. But ϕ is the phase and A is the magnitude, and neither has an effect on the period. So if $e^{j\theta_0 n}$ is periodic, then

$$e^{j\theta_0 n} = e^{j\theta_0(n+N)}$$

or

$$e^{j\theta_0 n} = e^{j\theta_0 n} e^{j\theta_0 N}$$

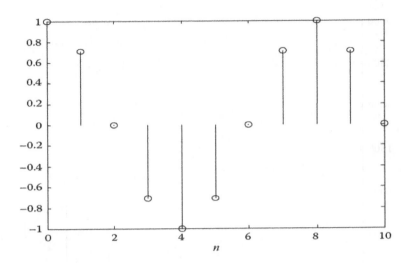

FIGURE 1.10   The sinusoidal discrete signal.

If we divide the previous equation by $e^{j\theta_0 n}$, we get

$$1 = e^{j\theta_0 N} = \cos(\theta_0 N) + j\sin(\theta_0 N)$$

For the previous equation to be true, the following two conditions must be true:

$$\cos(\theta_0 N) = 1$$

and

$$\sin(\theta_0 N) = 0$$

These two conditions can be satisfied only if $\theta_0 N$ is an integer multiple of 2π. In other words, $x(n)$ is periodic if

$$\theta_0 N = 2\pi k$$

where k is an integer. This can be written as

$$\frac{2\pi}{\theta_0} = \frac{N}{k}$$

If N/k is a rational number (ratio of two integers), then $x(n)$ is periodic and the period is

$$N = k\left(\frac{2\pi}{\theta_0}\right) \tag{1.7}$$

The smallest value of N that satisfies the previous equation is called the fundamental period. If $2\pi/\theta_0$ is not a rational number, then $x(n)$ is not periodic.

Example 1.2

Consider the following continuous signal for the current

$$i(t) = \cos(20\pi t)$$

which is sampled at 12.5 ms. Will the resulting discrete signal be periodic?

SOLUTION

The continuous radian frequency is $w = 20\pi$ rad. Since the sampling interval T_s is 12.5 ms = 0.0125 s,

$$x(n) = \cos\left(2\pi(10)(0.0125)n\right) = \cos\left(\frac{2\pi}{8}n\right) = \cos\left(\frac{\pi}{4}n\right)$$

Since for periodicity, we must have

$$\frac{2\pi}{\theta_0} = \frac{N}{k}$$

we get

$$\frac{2\pi}{2\pi/8} = \frac{N}{k} = \frac{16\pi}{2\pi} = \frac{8}{1}$$

For $k = 1$, we have $N = 8$, which is the fundamental period.

Example 1.3

Are the following discrete signals periodic? If so, what is the period for each?

1. $x(n) = 2\cos\left(\sqrt{2}\pi n\right)$
2. $x(n) = 20\cos\left(\pi n\right)$

SOLUTION

For the first signal, $\theta_0 = \sqrt{2}\pi$ and the ratio $2\pi/\theta_0$ must be a rational number.

$$\frac{2\pi}{\theta_0} = \frac{2\pi}{\sqrt{2}\pi} = \frac{2}{\sqrt{2}}$$

This is clearly not a rational number, and therefore, the signal is not periodic.

For the second signal, $\theta_0 = \pi$ and the ratio $2\pi/\theta_0 = 2\pi/\pi = 2$ is a rational number. Thus, the signal is periodic and the period is calculated by setting

$$2 = \frac{N}{k}$$

For $k = 1$, we get $N = 2$. Thus, N is the fundamental period.

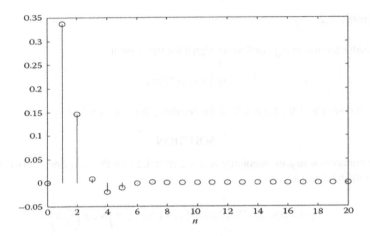

FIGURE 1.11   The decaying sinusoidal discrete signal.

1.9   EXPONENTIALLY MODULATED SINUSOIDAL SIGNAL

The exponentially modulated sinusoidal signal is written mathematically as

$$x(n) = A\alpha^n \cos(\theta_0 n + \phi) \tag{1.8}$$

If $\cos(\theta_0 n + \phi)$ is periodic and $0 < |\alpha| < 1$, $x(n)$ is a decaying exponential discrete signal, as shown in Figure 1.11. If $\cos(\theta_0 n + \phi)$ is periodic and $1 < |\alpha| < \infty$, $x(n)$ is a growing exponential discrete signal, as shown in Figure 1.12. If $\cos(\theta_0 n + \phi)$ is

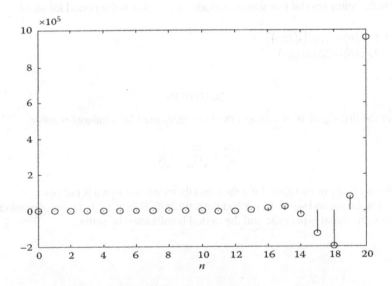

FIGURE 1.12   The growing sinusoidal discrete signal.

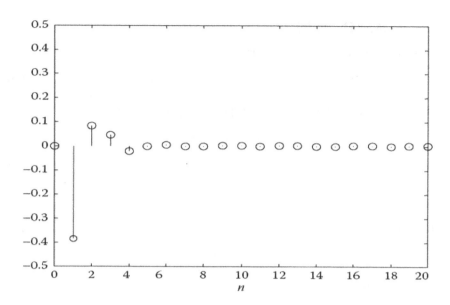

FIGURE 1.13   The irregularly decaying modulated sinusoidal discrete signal.

nonperiodic and $0 < |\alpha| < 1$, $x(n)$ will not decay exponentially in a regular fashion, as shown in Figure 1.13. If $\cos(\theta_0 n + \phi)$ is nonperiodic and $1 < |\alpha| < \infty$, $x(n)$ will grow irregularly without bounds, as shown in Figure 1.14.

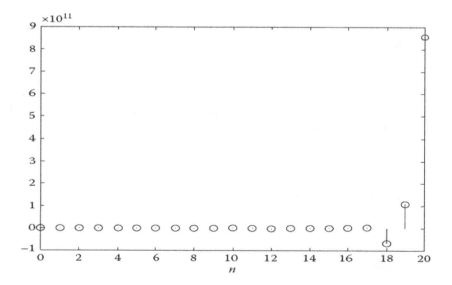

FIGURE 1.14   The irregularly growing modulated sinusoidal discrete signal.

1.10   COMPLEX PERIODIC DISCRETE SIGNAL

A complex discrete signal is represented mathematically as

$$x(n) = Ae^{j\alpha n} \quad -\infty < n < +\infty \tag{1.9}$$

where α is a real number. For $x(n)$ to be periodic we must have

$$x(n) = x(n + N)$$

or

$$Ae^{j\alpha n} = Ae^{j\alpha(n+N)}$$

Simplifying, we get

$$Ae^{j\alpha n} = Ae^{j\alpha n}e^{j\alpha N}$$

If we divide both sides by $Ae^{j\alpha n}$, we get

$$1 = e^{j\alpha N}$$

The previous equation is satisfied if

$$\alpha N = 2\pi k$$

or

$$\frac{N}{k} = \frac{2\pi}{\alpha}$$

Again, and similar to what we did in Section 1.8, if $2\pi/\alpha$ is a rational number, then $x(n)$ is periodic with period

$$N = \left(\frac{2\pi}{\alpha}\right)k$$

and the smallest N satisfying the previous equation is called the fundamental period. If $2\pi/\alpha$ is not rational, then $x(n)$ is not periodic.

Example 1.4

Consider the following complex sinusoidal discrete signals:

1. $2e^{jn}$
2. $e^{jn\pi}$
3. $e^{(j2\pi n + 2)}$
4. $e^{jn8\pi/3}$

Are the signals periodic? If so, what are their periods?

SOLUTION

For the first signal, $jn = j\alpha n$ requires that $\alpha = 1$. For periodicity, the ratio $2\pi/\alpha$ must be a rational number. But $2\pi/1$ is not a rational number and this signal is not periodic.

For the second signal, $jn\pi = j\alpha n$ requires $\alpha = \pi$. For periodicity again, $2\pi/\alpha$ must be a rational number. The ratio $2\pi/\alpha = 2\pi/\pi = 2/1$ is a rational number. Thus, the signal is periodic.

$$\frac{2\pi}{\alpha} = \frac{N}{k} = \frac{2\pi}{\pi} = \frac{2}{1}$$

For $k = 1$, we get $N = 2$. Thus, N is the fundamental period.

For the third signal, $e^{(j2\pi n + 2)}$ can be written as $e^2 e^{j2\pi n}$ and $2\pi j n = j\alpha n$ requires $\alpha = 2\pi$. For periodicity, $2\pi/\alpha = 2\pi/2\pi = 1/1$ must be a rational number, which is true in this case. Therefore, the signal is periodic.

$$\frac{2\pi}{\alpha} = \frac{2\pi}{2\pi} = \frac{1}{1} = \frac{N}{k}$$

For $k = 1$, we get $N = 1$, which is the fundamental period.

For the fourth signal, $j\alpha n = j8\pi/3$ and $\alpha = 8\pi/3$. For periodicity, $2\pi/\alpha$ must be a rational number.

$$\frac{2\pi}{\alpha} = \frac{2\pi}{8\pi/3} = \frac{6}{8} = \frac{3}{4}$$

This is a rational number and the signal is periodic with the fundamental period N calculated by setting

$$\frac{2\pi}{\alpha} = \frac{3}{4} = \frac{N}{k}$$

For $k = 4$ we get $N = 3$ as the smallest integer.

1.11   SHIFTING OPERATION

A shifted discrete signal $x(n)$ is $x(n - k)$, where k is an integer. If k is positive, then $x(n)$ is shifted k units to the right, and if k is negative, then $x(n)$ is shifted k units to the left.

Consider the discrete impulse signal $x(n) = 5A\delta(n)$. The signal $x(n - 1) = 5\delta(n - 1)$ is the signal $x(n)$ shifted by 1 unit to the right. Also $x(n + 3) = 5\delta(n + 3)$ is $x(n)$ shifted by 3 units to the left. The importance of the shift operation will be apparent when we get to the next chapters. In the following section, we will see one basic importance.

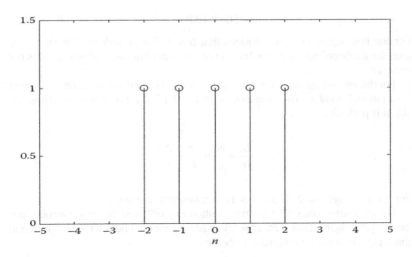

FIGURE 1.15   Pulse signal for Example 1.5.

Example 1.5

Consider the discrete pulse in Figure 1.15. Write this pulse as a sum of discrete step signals.

SOLUTION

Consider the shifted step signal in Figure 1.16 and the other shifted step signal in Figure 1.17. Let us subtract Figure 1.16 from Figure 1.17 to get Figure 1.15.

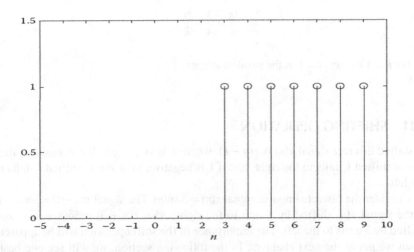

FIGURE 1.16   Signal for Example 1.5.

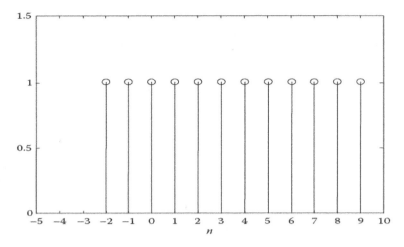

FIGURE 1.17 Signal for Example 1.5.

1.12 REPRESENTING A DISCRETE SIGNAL USING IMPULSES

Any discrete signal can be represented as a sum of shifted impulses. Consider the signal in Figure 1.18 that has values at $n = 0, 1, 2$, and 3.

Each of these values can be thought of as an impulse shifted by some units. The signal at $n = 0$ can be represented as $1(n)$, the signal at $n = 1$ as $1.5(n - 1)$, the signal at $n = 2$ as $1/2(n - 2)$, and the signal at $n = 3$ as $1/4(n - 3)$. Therefore, $x(n)$ in Figure 1.18 can be represented mathematically as

$$x(n) = \delta(n) + 1.5\delta(n-1) + \frac{1}{2}\delta(n-2) + \frac{1}{4}\delta(n-3)$$

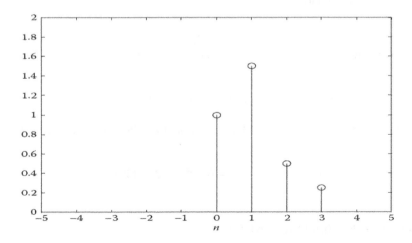

FIGURE 1.18 Representation of a discrete signal using impulses.

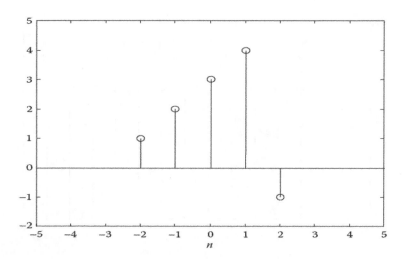

FIGURE 1.19   Signal for Example 1.6.

Example 1.6

Represent the following discrete signals using impulse signals.

1. $x(n) = \left\{\begin{array}{ccccc} 1, & 2, & 3, & 4, & -1 \\ & & \uparrow & & \end{array}\right\}$

2. $x(n) = \left\{\begin{array}{cccc} 0, & 1, & 2, & -4 \\ & \uparrow & & \end{array}\right\}$

where the arrow under the number indicates $n = 0$, where n is the time index where the signal starts.

SOLUTION

Graphically, the first signal is shown in Figure 1.19, and it can be seen as a sum of impulses as

$$x(n) = 1\delta(n+2) + 2\delta(n+1) + 3\delta(n) + 4\delta(n-1) - 1\delta(n-2)$$

The second signal is shown in Figure 1.20 and can be written as the sum of impulses as

$$x(n) = 0\delta(n) + 1\delta(n-1) + 2\delta(n-2) - 4\delta(n-3)$$

1.13   REFLECTION OPERATION

Mathematically, a reflected signal $x(n)$ is written as $x(-n)$, as shown in Figure 1.21.

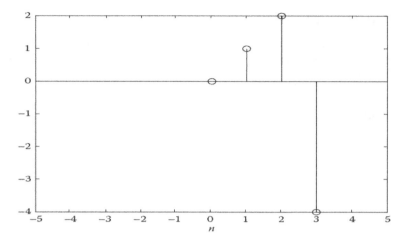

FIGURE 1.20 Signal for Example 1.6.

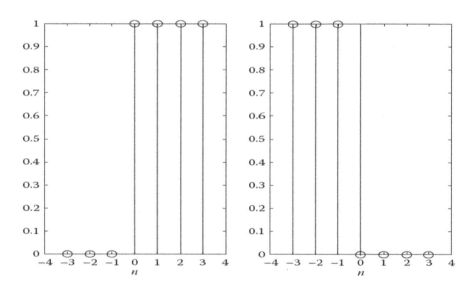

FIGURE 1.21 The reflected signal.

1.14 TIME SCALING

A time-scaled discrete signal $x(an)$ of $x(n)$ is calculated by considering two cases for a. In case $a = k$, we will consider all cases where k is an integer. In case $a = 1/k$, we will also consider all values for k where k is an integer. An example will be given shortly.

1.15   AMPLITUDE SCALING

An amplitude-scaled version of $x(n)$ is $ax(n)$ where, if a is negative, the magnitude of each sample in $x(n)$ is reversed and scaled by the absolute value of a. If a is positive, then each sample of $x(n)$ is scaled by a.

Example 1.7

Consider the following signal:

$$x(n) = \left\{ \begin{matrix} -1,2,0,3 \\ \uparrow \end{matrix} \right\}$$

Find

1. $x(-n)$
2. $x(-n + 1)$
3. $2x(-n + 1)$
4. $x(-n) + x(-n + 1)$

SOLUTION

1. $x(n) = \left\{ \begin{matrix} -1,2,0,3 \\ \uparrow \end{matrix} \right\}$

   $x(-n)$ is $x(n)$ shifted about the zero position, which is

   $$x(-n) = \left\{ \begin{matrix} 3,0,2,-1 \\ \uparrow \end{matrix} \right\}$$

2. $x(-n + 1)$ is $x(-n)$ shifted to the left 1 unit and it is

   $$x(-n+1) = \left\{ \begin{matrix} 3,0,2,-1 \\ \uparrow \end{matrix} \right\}$$

3. $2x(-n + 1)$ is $x(-n + 1)$ scaled by 2 and is

   $$2x(-n+1) = \left\{ \begin{matrix} 6,0,4,-2 \\ \uparrow \end{matrix} \right\}$$

4. $x(-n) + x(-n+1) = \left\{ \begin{matrix} 3,0,2,-1 \\ \uparrow \end{matrix} \right\} + \left\{ \begin{matrix} 3,0,2,-1 \\ \uparrow \end{matrix} \right\}$

   We add these two signals at the $n = 0$ index to get

   $$x(-n) + x(-n+1) = \left\{ \begin{matrix} 0,3,0,2,-1 \\ \uparrow \end{matrix} \right\} + \left\{ \begin{matrix} 3,0,2,-1,0 \\ \uparrow \end{matrix} \right\} = \left\{ \begin{matrix} 3,3,2,1,-1 \\ \uparrow \end{matrix} \right\}$$

1.16   EVEN AND ODD DISCRETE SIGNAL

Every discrete signal $x(n)$ can be represented as a sum of an odd and an even discrete signal. The discrete signal $x(n)$ is odd if

$$x(n) = -x(n) \tag{1.10}$$

The discrete signal $x(n)$ is even if

$$x(n) = x(-n) \tag{1.11}$$

Therefore, any discrete signal, $x(n)$, can be written as the sum of an even and an odd discrete signal. We write

$$x(n) = x_{\text{even}}(n) + x_{\text{odd}}(n)$$

where

$$x_{\text{even}}(n) = \frac{1}{2}(x(n) + x(-n)) \tag{1.12}$$

and

$$x_{\text{odd}}(n) = \frac{1}{2}(x(n) - x(-n)) \tag{1.13}$$

Example 1.8

The signals in Figures 1.22 and 1.23 are not odd or even. Write them as the sum of even and odd signals.

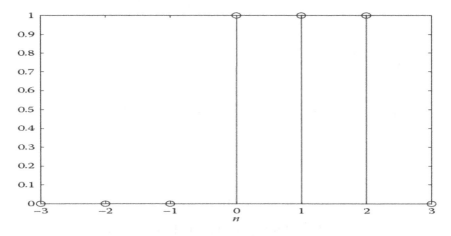

FIGURE 1.22   Signal for Example 1.8.

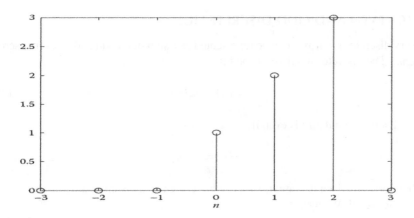

FIGURE 1.23   Signal for Example 1.8.

SOLUTION

For $x(n) = \left\{ \underset{\uparrow}{1}, 1, 1 \right\}$, we have

$$x_{odd}(n) = \frac{1}{2}\big[x(n) - x(-n)\big] = \frac{1}{2}\left[\left\{\underset{\uparrow}{1}, 1, 1\right\} - \left\{1, 1, \underset{\uparrow}{1}\right\}\right] = \left\{-1/2, -1/2, \underset{\uparrow}{0}, 1/2, 1/2\right\}$$

$$x_{even}(n) = \frac{1}{2}\big[x(n) + x(-n)\big] = \frac{1}{2}\left[\left\{\underset{\uparrow}{1}, 1, 1\right\} + \left\{1, 1, \underset{\uparrow}{1}\right\}\right] = \left\{1/2, 1/2, \underset{\uparrow}{1}, 1/2, 1/2\right\}$$

Now if we add x_{odd} and x_{even}, we will get

$$\left\{-1/2, -1/2, \underset{\uparrow}{0}, 1/2, 1/2\right\} + \left\{1/2, 1/2, \underset{\uparrow}{1}, 1/2, 1/2\right\} = \left\{0, 0, \underset{\uparrow}{1}, 1, 1\right\}$$

For $x(n) = \left\{\underset{\uparrow}{1}, 2, 3\right\}$, we have

$$x_{odd}(n) = \frac{1}{2}\big[x(n) - x(-n)\big] = \frac{1}{2}\left[\left\{\underset{\uparrow}{1}, 2, 3\right\} - \left\{3, 2, \underset{\uparrow}{1}\right\}\right] = \left\{-3/2, -1, \underset{\uparrow}{0}, 1, 3/2\right\}$$

$$x_{even}(n) = \frac{1}{2}\big[x(n) + x(-n)\big] = \frac{1}{2}\left[\left\{\underset{\uparrow}{1}, 2, 3\right\} + \left\{3, 2, \underset{\uparrow}{1}\right\}\right] = \left\{3/2, 1, \underset{\uparrow}{0}, 1, 3/2\right\}$$

Now if we add x_{odd} and x_{even}, we will get

$$\left\{-3/2, -1, \underset{\uparrow}{0}, 1, 3/2\right\} + \left\{3/2, 1, \underset{\uparrow}{0}, 1, 3/2\right\} = \left\{0, 0, \underset{\uparrow}{0}, 2, 3\right\}$$

1.17   DOES A DISCRETE SIGNAL HAVE A TIME CONSTANT?

Consider the continuous real exponential signal

$$x(t) = e^{-\alpha t}$$

Let us sample $x(t)$ every $t = nT_s$ s with $a > 0$. Then we write

$$x(n) = e^{-\alpha n T_s} = \left(e^{-\alpha T_s}\right)^n = a^n$$

where $a = e^{-\alpha T_s}$ is a real number. The continuous exponential signal has a time constant τ, where $e^{-\alpha t} = e^{-t/\tau}$. This implies that $\tau = 1/a$. Therefore,

$$x(n) = \left(e^{-T_s/\tau}\right)^n = a^n$$

But $e^{-T_s/\tau} = a$ and by taking the logarithm on both sides, we get

$$\frac{\tau}{T_s} = -\frac{1}{\ln a} \quad \text{or} \quad \tau = \frac{-T_s}{\ln a}$$

So, a discrete time signal that was obtained by sampling a continuous exponential signal has a time constant as given earlier.

Example 1.9

Consider the following two continuous signals:

$$x(t) = e^{-t}$$
$$x(t) = e^{-100t}$$

1. Find the time constant for both continuous signals.
2. Discretize both signals for $T_s = 1$ s and then find the time constant for the resulting discrete signals.

SOLUTION

For $x(t) = e^{-t}$ and by setting $e^{-t} = e^{-t/\tau}$ in order to find the time constant τ we see that $\tau = 1$ s. By letting $t = nT_s$ we get the discretized signal

$$x(n) = e^{-(nT_s)} = e^{-n} = \left(e^{-1}\right)^n$$

As discussed earlier, we set

$$e^{-1} = e^{-T_s/\tau}$$

to get the discrete time constant

$$\tau = \frac{-T_s}{\ln\left(e^{-1}\right)} = \frac{-1}{\ln\left(e^{-1}\right)}$$

For $x(t) = e^{-100t}$, we set

$$e^{-100t} = e^{-t/\tau}$$

to get $\tau = 1/100$ s. Similarly, the second discretized signal is

$$x(n) = e^{-100T_s n} = \left(e^{-100}\right)^n$$

and the discrete time constant is

$$\tau = \frac{-T_s}{\ln\left(e^{-100}\right)} = \frac{-1}{\ln\left(e^{-100}\right)}$$

1.18   BASIC OPERATIONS ON DISCRETE SIGNALS

A discrete system will operate on discrete inputs. Some basic operations follow. The output of a discrete system is the input operated upon in a certain way.

1.18.1   MODULATION

Consider the two discrete signals $x_1(n)$ and $x_2(n)$. The resulting signal $y(n)$ where

$$y(n) = x_1(n)x_2(n)$$

is called the modulation of $x_1(n)$ and $x_2(n)$. $y(n)$ is a discrete signal found by multiplying the sample values of $x_1(n)$ and $x_2(n)$ at every instant.

1.18.2   ADDITION AND SUBTRACTION

Consider the two discrete signals $x_1(n)$ and $x_2(n)$. The addition/subtraction of these two signals is $y(n)$ where

$$y(n) = x_1(n) \pm x_2(n)$$

1.18.3   SCALAR MULTIPLICATION

A scalar multiplication of the discrete signal $x(n)$ is the signal $y(n)$ where

$$y(n) = Ax(n)$$

where A is the scaling factor.

1.18.4 COMBINED OPERATIONS

We may have multiple operations among input discrete signals. One operation may be represented as

$$y(n) = 2x_1(n) + 1.5x_2(n-1)$$

where we have scaling, shifting, and addition operations combined. In other operations you may have the output $y(n)$ presented as

$$y(n) = 2x_1(n) - x_2(n)x_3(n)$$

where you have $x_1(n)$ scaled, then the modulated signal resulting from $x_2(n)$ and $x_3(n)$ is subtracted from $2x_1(n)$.

Example 1.10

Consider the discrete signal as shown in Figure 1.24. Find

1. $x(-n)$
2. $x(n+2)$
3. $x(2n)$
4. $x(n/2)$
5. $x(2n-1)$
6. $x(n)x(n)$

SOLUTION

From Figure 1.24, we see that

$$x(n) = \left\{ 1,1,\underset{\uparrow}{0}, -1, -1 \right\}$$

1. $x(-n)$ is $x(n)$ reflected and is

$$x(-n) = \left\{ -1, -1, \underset{\uparrow}{0}, 1, 1 \right\}$$

2. $x(n+2)$ is $x(n)$ shifted to the left by 2 units and is

$$x(n+2) = \left\{ 1,1,0,-1,\underset{\uparrow}{-1} \right\}$$

3. The new time axis in this case is $n_{new} = n_{old}/2$, where n_{old} is the index of the given $x(n)$. The indices are arranged as in the following with the help of the equation $n_{new} = n_{old}/2$.

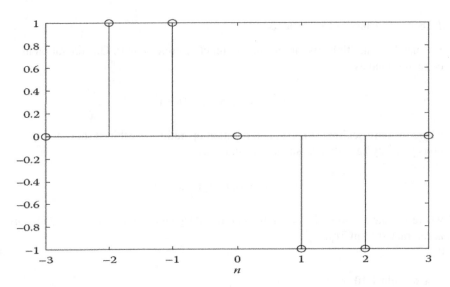

FIGURE 1.24 Signal for Example 1.10.

n_{old} n_{new}
-2   -1
-1   -1/2 No value will appear at $n = -1/2$ since n must be an integer
 0   0
 1   1/2 No value will appear
 2   1
 3   No value for $x(n)$ at $n = 3$, so we stop at this point

Therefore, the new n axis for $x(2n)$ will start at –1 and will contain the indices $n = -1$, 0, and 1. The value at –1 for the scaled signal $x(2n)$ will be the value of $x(n)$ at $n = -2$. Similarly, the value at $n_{new} = 0$ will be the value at $n_{old} = 0$, and the value at $n_{new} = 1$ will be the value at $n_{old} = 2$. Therefore, $x(2n)$ is

$$x(2n) = \left\{ \begin{matrix} 0,1,0,-1,0 \\ \uparrow \end{matrix} \right\}$$

4. This is also a scaling case. For this case, the new time axis is

$$n_{new} = 2n_{old}$$

These indices are arranged in the following table:

n_{old} n_{new}
-2   -4
-1   -2
 0   0
 1   2
 2   4

The scaled signal $x(n/2)$ therefore is

$$x\left(\frac{n}{2}\right) = \left\{\begin{matrix} 1,0,1,0,0,0,-1,0,-1 \\ \uparrow \end{matrix}\right\}$$

5. $x(2n + 1)$ is $x(2n)$ shifted left by one unit and is

$$x(2n+1) = \left\{\begin{matrix} 0,1,0,-1,0 \\ \uparrow \end{matrix}\right\}$$

6. $x(n)$ multiplied by $x(n)$ is called the element-by-element multiplication and is

$$x(n)x(n) = \left\{\begin{matrix} 1,1,0,-1,-1 \\ \uparrow \end{matrix}\right\}\left\{\begin{matrix} 1,1,0,-1,-1 \\ \uparrow \end{matrix}\right\} = \left\{\begin{matrix} 1,1,0,1,1 \\ \uparrow \end{matrix}\right\}$$

1.19   ENERGY AND POWER DISCRETE SIGNALS

The total energy in the discrete signal $x(n)$ is E and mathematically written as

$$E = \sum_{-\infty}^{+\infty} |x(n)|^2 \tag{1.14}$$

The average power in the discrete signal $x(n)$ is P and is written mathematically as

$$P = \lim_{k \to \infty} \frac{1}{2k+1} \sum_{n=-k}^{k} |x(n)|^2 \tag{1.15}$$

where

$$\sum_{n=-k}^{n=k} |x(n)|^2$$

is the energy in $x(n)$ in the interval $-k < n < k$. The average power of a discrete periodic signal is written mathematically as

$$P = \frac{1}{N} \sum_{n=0}^{N-1} |x(n)|^2 \tag{1.16}$$

where N is the period of $x(n)$.

Example 1.11

Consider the following finite discrete signals:

1. $x(n) = -1\delta(n - 0) + 2\delta(n - 1) - 2\delta(n - 2)$
2. $x(n) = \left\{ \begin{matrix} 1, 0, -1 \\ \uparrow \end{matrix} \right\}$

Find the energy in both signals.

SOLUTION

The first signal $x(n)$ can be written as

$$x(n) = \left\{ \begin{matrix} -1, 2, -2 \\ \uparrow \end{matrix} \right\}$$

The energy in the signal is then

$$E = \sum_{n=0}^{2} \left| x(n)^2 \right| = (-1)^2 + (2)^2 + (-2)^2 = 1 + 4 + 4 = 9$$

This means that $x(n)$ has finite energy.

For the second signal the total energy is given by

$$E = \sum_{n=-1}^{1} \left| x(n)^2 \right| = (1)^2 + (0)^2 + (-1)^2 = 2$$

Example 1.12

Consider the discrete signals

$$x(n) = 2(-1)^n \quad n \geq 0$$

Find the energy and the power in $x(n)$.

SOLUTION

$$E = \sum_{n=0}^{\infty} \left| 2(-1)^2 \right|^2 = 4[1 + 1 + \cdots]$$

This sum clearly does not converge to a real number. Hence, $x(n)$ has infinite energy. The power in the signal is

$$P = \lim_{k \to \infty} \frac{1}{2k+1}\left(4\sum_{n=0}^{k}1\right) = \lim_{k \to \infty} \frac{4(k+1)}{2k+1} = 2$$

Therefore, $x(n)$ has finite power and is a power signal.

Example 1.13

Find the energy in the signal:

$$x(n) = \frac{1}{n} \qquad n \geq 1$$

SOLUTION

The energy is given by

$$E = \sum_{n=1}^{\infty}\left|\frac{1}{n}\right|^2 = 1 + \left(\frac{1}{2}\right)^2 + \left(\frac{1}{3}\right)^2 + \cdots$$

which converges to $\pi^2/6$. This means that $x(n)$ has finite energy.

1.20   BOUNDED AND UNBOUNDED DISCRETE SIGNALS

A discrete signal $x(n)$ is bounded if each sample in the signal has a bounded magnitude. Mathematically, if $x(n)$ is bounded then

$$|x(n)| \leq \beta < \infty$$

where β is some positive value. If this is not the case, then $x(n)$ is said to be unbounded. The step signal is bounded, the ramp is unbounded, and the sinusoidal signal is bounded.

1.21   SOME INSIGHTS: SIGNALS IN THE REAL WORLD

The signals that we have introduced in this chapter were all represented in mathematical form and plotted on graphs. This is how we represent signals for the purpose of analysis, synthesis, and design. For better understanding of these signals, we will provide herein real-life situations and see the relation between these mathematical abstractions of signals and get a feel of what they may represent in real life.

1.21.1 STEP SIGNAL

In real-life situations, this signal can be viewed as a constant force of magnitude A newtons applied at time $(t) = 0$ s to a certain object for a long time. In another situation, $Au(t)$ can be an applied voltage of constant magnitude to a load resistor R at the time $t = 0$.

1.21.2 IMPULSE SIGNAL

Again, in real life, this signal can represent a situation in which a person hits an object with a hammer with a force of A newtons for a very short period of time (picoseconds). We sometimes refer to this kind of signal as a shock.

In another real-life situation, the impulse signal can be as simple as closing and opening an electrical switch in a very short time. Another situation where a spring-carrying mass is hit upward can be seen as an impulsive force. A sudden oil spill similar to the one that happened during the Gulf war can represent a sudden flow of oil. You may realize that it is impossible to generate a pure impulse signal for zero duration and infinite magnitude. To create an approximation to an impulse, we can generate a pulse signal of very short duration, where the duration of the pulse signal is very short compared with the response of the system.

1.21.3 SINUSOIDAL SIGNAL

This signal can be thought of as a situation where a person is shaking an object regularly. This is like pushing and pulling an object continuously with a period of T s. Thus, a push and pull forms a complete period of shaking. The distance the object covers during this shaking represents a sinusoidal signal. In the case of electrical signals, an AC voltage source is a sinusoidal signal.

1.21.4 RAMP SIGNAL

In real-life situations, this signal can be viewed as a signal that is increasing linearly with time. An example is when a person applies a force at time $t = 0$ to an object and keeps pressing the object with increasing force for a long time. The rate of the increase in the force applied is constant.

Consider another situation where a radar antenna, an antiaircraft gun, and an incoming jet are in one place. The radar antenna can provide an angular position input. In one case, the jet motion forces this angle to change uniformly with time. This will force a ramp input signal to the antiaircraft gun since it will have to track the jet.

1.21.5 OTHER SIGNALS

A train of impulses can be thought of as hitting an object with a hammer continuously and uniformly. In terms of electricity, you may be closing and opening a switch continuously. A rectangular pulse can be likened to applying a constant force to an object for a certain time and instantaneously removing that force. It is also like applying a constant voltage for a certain time and then instantaneously closing the switch of the voltage source. Other signals are the random signals where the magnitude changes randomly as time progresses. These signals can be thought of as shaking an object with variable random force as time progresses, or as a gusting wind.

1.22   END-OF-CHAPTER EXAMPLES

EOCE 1.1

Consider the following discrete periodic and nonperiodic signals in Figure 1.25. For the periodic signal, find the period N and also the average power. For the nonperiodic signals, find the total energy.

SOLUTION

For the first discrete signal in Figure 1.25, the signal $x(n)$ is periodic with period $N = 3$. The average power is then

$$P = \frac{1}{N}\sum_{n=0}^{N-1}|x(n)|^2 = \frac{1}{3}\sum_{n=0}^{2}|x(n)|^2 = \frac{1}{3}\left((1)^2 + \left(\frac{1}{2}\right)^2 + (0^2)\right)$$

$$P = \frac{1}{3}\left(1 + \frac{1}{4}\right) = \frac{1}{3}\left(\frac{4+1}{4}\right) = \frac{5}{12}$$

We can use MATLAB® to calculate this average power as in the script EOCE1_11. The result will be 0.4167.

For the second discrete signal the period is $N = 5$. The average power is calculated as

$$P = \frac{1}{N}\sum_{n=0}^{N-1}|x(n)|^2 = \frac{1}{5}\sum_{n=0}^{3}|x(n)|^2 = \frac{1}{5}\left((x(0))^2 + (x(1))^2 + (x(2))^2 + (x(3))^2\right)$$

$$P = \frac{1}{5}\left((1)^2 + (1)^2 + (1)^2 + (1)^2 + (0)^2\right) = \frac{1}{5}[4] = 0.8$$

We can use MATLAB and write the script EOCE1_12 to find the power. The result is 0.8 for the average power.

The last signal in Figure 1.25 is not periodic and the total energy is

$$E = \sum_{n=0}^{+\infty}|x(n)|^2 = \sum_{n=0}^{N=4}|x(n)|^2 = x(0)^2 + x(1)^2 + x(2)^2 + x(3)^2 + x(4)^2$$

$$E = (0)^2 + (1)^2 + (2)^2 + (2)^2 = 1 + 4 + 4 + 4 = 13$$

We can use MATLAB to find this total energy in the signal by writing the EOCE1_13 script. The result will be 13 for the energy.

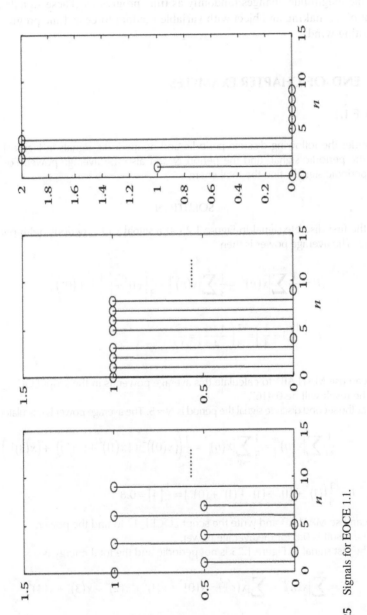

FIGURE 1.25 Signals for EOCE 1.1.

EOCE 1.2

Write MATLAB scripts to simulate the step and the impulse signals.

SOLUTION

For the step signal,

$$u(n) = \begin{cases} 1 & n \geq 0 \\ 0 & n < 0 \end{cases}$$

We can use the MATLAB function ones to generate sequences that have every value unity. We write ones $(1, L)$ where L is the number of ones in this row vector that "ones" generate. However, sometimes not all step signals start at $n = 0$. In this case, we have

$$u(n - n_0) = \begin{cases} 1 & n \geq n_0 \\ 0 & n < 0 \end{cases}$$

We know that $u(n)$ is defined for $n \geq 0$ and we also know that we cannot generate $u(n)$ for an infinite number of samples. Therefore, we will generate these sequences for a limited interval. We will denote n_1 to be the left limit and n_2 to be the right limit. Notice that ones $(1, L)$ will generate L ones for $n = 0, 1, 2, 3, \ldots, L-1$ automatically. To generate the most general step sequence $u(n - n_0)$ in the interval $n_1 \leq n \leq n_2$, we write the MATLAB script EOCE1_21 that will generate a unit step signal that starts at $n_0 = -1$ and defined in the interval $-3 \leq n \leq 3$ first.

Note in the previous script on the line before the last that for $n = -3$, $n_0 = -1$ (fixed), the logical expression $(-3 - (-1)) \geq 0$ results in $-3 + 1 \geq 0$ and evaluates to false, which in MATLAB evaluates to zero. Therefore, for $n = -3$, $u(n + 1)$ is zero.

Now take $n = -2$; the expression $-2 + 1 \geq 0$ evaluates to false again. Take $n = -1$. The expression $-1 + 1 \geq 0$ evaluates to true, and hence, the first one appears. The plot is shown in Figure 1.26. To sketch $3u(n - 5)$ in the interval $-10 \leq n \leq 10$, we write the script EOCE1_22 and the plot is shown in Figure 1.27.

For the impulse signal, we have

$$\delta(n) = \begin{cases} 1 & n = 0 \\ 0 & n \neq 0 \end{cases}$$

We can use the MATLAB function zeros to generate a sequence of zeros of any length. For example, the command

```
delta = zeros (1, L)
```

will generate a sequence of zeros of length L. Then we can make the first value in the sequence one to form the impulse signal and write

```
delta (1) = 1;
```

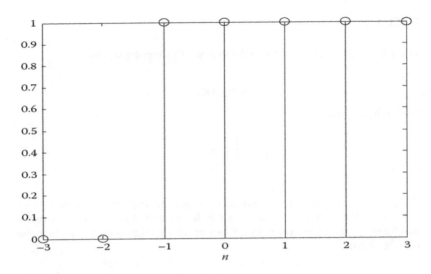

FIGURE 1.26   MATLAB®-generated step signal starting at $n = -1$.

But again, suppose you want to generate an impulse sequence that has a certain value at $n = n_0$ and zero otherwise. We write

$$\delta(n - n_0) = \begin{cases} 1 & n = n_0 \\ 0 & n \neq n_0 \end{cases}$$

For this, we write the MATLAB script EOCE1_23 that simulates $\delta(n - 1)$.

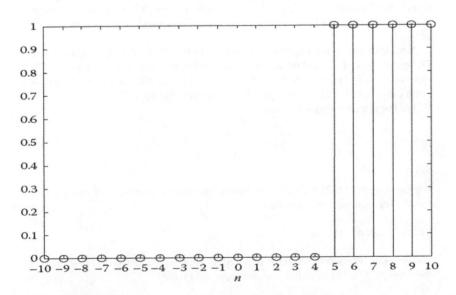

FIGURE 1.27   MATLAB®-generated step signal starting at $n = 5$.

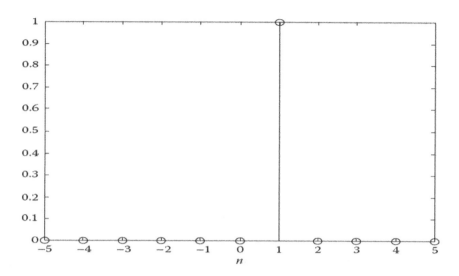

FIGURE 1.28 MATLAB®-generated impulse signal at $n = 1$.

The plot is shown in Figure 1.28. Note the last line of the previous MATLAB script.

$$x = \left[(n - n_0) == 0 \right]$$

If $n = -5$, $(-5 - 1 == 0)$ evaluates to false and the impulse at $n = -5$ is zero.
If $n = 2$, $(2 - 1 == 0)$ also evaluates to zero and the impulse signal at $n = 2$ is zero. But if $n = 1$, $(1 - 1 == 0)$ evaluates to true and this is the only value in the interval $-5 \leq n \leq 5$ for the impulse to have a value other than zero.

EOCE 1.3

Write MATLAB scripts to simulate the exponential signal

$$x(n) = A(\alpha)^n$$

and the sinusoidal signal

$$x(n) = A\cos(\theta_0 n + \phi)$$

SOLUTION

For

$$x(n) = A(\alpha)^n \quad -\infty < n < +\infty$$

we can generate the $x(n)$ sequence in a limited interval only. To simulate $3(.5)^n$ in the interval $-3 \leq n \leq 3$, we write the EOCE1_31 script and the plot is shown in Figure 1.29. It is seen that this signal is bounded because $0 < \alpha < 1$.

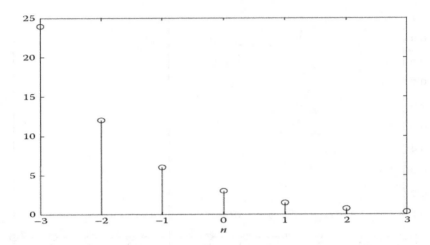

FIGURE 1.29   MATLAB®-generated exponential decaying signal.

For the sinusoid sequence

$$x(n) = A\cos(\theta_0 n + \varphi) \qquad -\infty < n < +\infty$$

the signal can be simulated in a fixed interval. To do that, let us look at the signal

$$x(n) = 3\cos(3\pi n + 5) \qquad -10 < n < 10$$

and write the EOCE1_32 script to simulate this signal.
    The plot is shown in Figure 1.30.

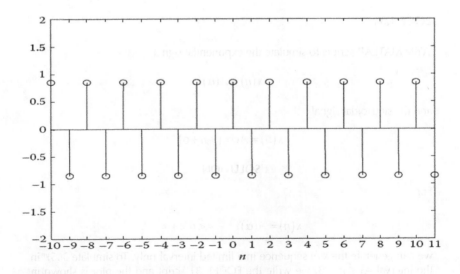

FIGURE 1.30   MATLAB®-generated sinusoidal signal.

EOCE 1.4

Consider the following signals:

$$x(n) = (.5)^n \cos(2n\pi + \pi)$$

$$x(n) = 5\cos(2n\pi + \pi) + 3$$

Are the signals periodic?

SOLUTION

The first signal will decay to zeros as the index n get larger and it is not periodic. For the second signal, $\theta_0 = 2\pi$ and for periodicity, the ratio $2\pi/\theta_0$ must be rational. We have

$$\frac{2\pi}{\theta_0} = \frac{2\pi}{2\pi} = \frac{1}{1}$$

and therefore, the signal is periodic. The period N is

$$N = k\left(\frac{2\pi}{\theta_0}\right)$$

For $k = 1$ we have $N = 1$. Note that the addition of 3 to $x(n)$ has no effect on the period N. We can use MATLAB to verify this and simulate $x(n)$ in the interval $-3 < n < 3$ and write the EOCE1_4 script and the plot is seen in Figure 1.31. It is seen from the figure that $N = 1$.

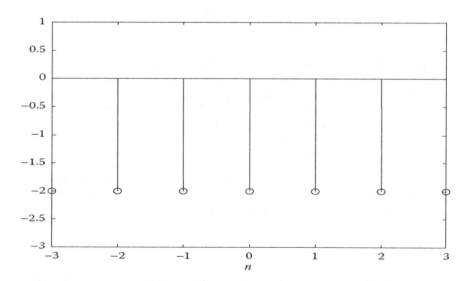

FIGURE 1.31 Signal for EOCE 1.4.

EOCE 1.5

Consider the following discrete signals:

$$x_1(n) = \left\{ \underset{\uparrow}{0}, 1, 2, 3 \right\}$$

$$x_2(n) = \left\{ 0, \underset{\uparrow}{1}, 2, 3 \right\}$$

Find $x_1(n) + x_2(n)$ and $x_1(n) \, x_2(n)$ analytically and using MATLAB.

SOLUTION

Analytically, we can arrange the two signals as follows and then add the corresponding samples. Remember we add samples with similar indices.

$$x_1(n) = \left\{ 0, \underset{\uparrow}{0}, 1, 2, 3 \right\} + x_2(n) = \left\{ 0, \underset{\uparrow}{1}, 2, 3, 0 \right\} = x(n) \left\{ 0, \underset{\uparrow}{1}, 3, 5, 3 \right\}$$

$x_1(n)$ has an index that starts at $n = 0$ and ends at $n = 3$, and $x_2(n)$ has an index that starts at $n = -1$ and ends at $n = 2$. Note also that $x(n)$ starts at $n = -1$ and ends at $n = 3$. The initial index, $n = -1$, for $x(n)$ is the minimum of the minimum of the starting index for $x_1(n)$ and $x_2(n)$.

The last index $n = 3$ for $x(n)$ is the maximum of the maximum of the last index for $x_1(n)$ and $x_2(n)$. This note will help us do the addition with MATLAB.

For $x_1(n) \, x_2(n)$, we have the same arrangement.

$$x(n) = \left\{ 0, \underset{\uparrow}{0}, 1, 2, 3 \right\} \left\{ 0, \underset{\uparrow}{1}, 2, 3, 0 \right\} = \left\{ 0, \underset{\uparrow}{0}, 2, 6, 0 \right\}$$

Using MATLAB, let n_1 represent the timescale (index) for $x_1(n)$ and n_2 represent the timescale for $x_2(n)$. Then the index for $x(n) = x_1(n) + x_2(n)$ will start with the minimum of n_1 and n_2 and end with the maximum of n_1 and n_2. Notice also when we added the signals analytically, we made both of the same length and the same length as $x(n)$. In MATLAB, the function finds works as follows: find(x) returns the indices of the vector x that are nonzero. With that notice we now write the script EOCE1_51 that adds the two signals.

The output is similar to what we got earlier.

Using MATLAB, we can simulate the element-by-element multiplication using a script similar to the one we just wrote for the addition. We will use EOCE1_52 and the result is again similar to what we arrived at before.

EOCE 1.6

The scripts that we have generated can be put in the form of functions. A function in MATLAB receives parameters and sends back results. Once this function is written and saved, it can be utilized as often as desired. The function is typed in the MATLAB editor and then saved and given a name similar to the name of the function that was written. Let us write functions to implement the step signal, the impulse signal, the reflection of a signal, the sum of two signals, the product of two signals, and the shifting by n_0 of a discrete signal.

SOLUTION

The general form of a MATLAB function is

```
Function [rv1 rv2 … rvn] = Function_Name (pv1, pv2, … pvn)
```

where

rv1 is the returned value one
pv1 is the passed value one

Function_Name is the name of the function, which should be the same name given to the file when the function is saved.

For the step discrete signal, let us call the function stepsignal. We will pass to stepsignal the time when the signal should start. We will call this time Sindex. We will also pass to it the starting and the ending of the time interval, which we will call Lindex and Rindex for left index and right index. The function will return the signal $x(n)$ and its index. The function is

```
function [xofn, index] = stepsignal (Sindex, Lindex, Rindex)
index = [Lindex: Rindex];
xofn = [(index - Sindex) > = 0];
```

For the impulse signal, let us call the function impulsesignal. We will pass to it the point of application of the impulse signal, Sindex, and the range, Lindex and Rindex. The function will send to us the impulse signal $x(n)$ and its index. The function is

```
function [xofn, index] = impulsesignal (Sindex, Lindex, Rindex)
index = [Lindex: Rindex];
xofn = [(index - Sindex) = = 0];
```

The reflection of the signal $x(n)$ is implemented using the MATLAB function fliplr. We will use fliplr to flip the sample values for $x(n)$ and to flip the time index. The function will be called xreflected and will receive the original $x(n)$ and the original index. It will give back the reflected $x(n)$ and the new index. The function is

```
function [xnew, nnew] = xreflected (xold, nold);
xnew = fliplr (xold);
nnew = -fliplr (nold);
```

The function to add two discrete signals $x_1(n)$ and $x_2(n)$ will be called x1plusx2. It will receive the original signals and their original indices and return the sum of the two signals and the index for the sum. The function to multiply $x_1(n)$ and $x_2(n)$ is similar to the x1plusx2 and is called x1timesx2.

A shifted version of $x(n)$ is $x(n - n_0)$

$$x_{\text{shift}}(n) = x(n - n_0)$$

If $n - n_0 = m$, then $x_{\text{shift}}(m + n_0) = x(m)$. This indicates that the sample values are not affected, but the index is changed by adding the index shift n_0. We will call the function xshifted and pass to it the original $x(n)$, the original index n_1, and the amount of shift n_0.

EOCE 1.7

Find

    1. $x(n) = u(n) - 3\delta(n - 1)$ $-3 \leq n \leq 3$
    2. $x(n) = 3u(n - 3) + \delta(n - 2) + u(-n)$ $-3 \leq n \leq 3$

SOLUTION

For the first $x(n)$, we have two signals as follows:

$$u(n) = \left\{ \begin{matrix} 0,0,0,1,1,1,1 \\ \uparrow \end{matrix} \right\}$$

and

$$-3\delta(n - 1) = \left\{ \begin{matrix} 0,0,0,0,-3,0,0 \\ \uparrow \end{matrix} \right\}$$

Thus, $u(n) - 3\delta(n - 1)$ is

$$x(n) = \left\{ \begin{matrix} 0,0,0,1,-2,1,1 \\ \uparrow \end{matrix} \right\}$$

We can use MATLAB to generate $x(n)$ and use the script EOCE1_71 and the plot is shown in Figure 1.32.

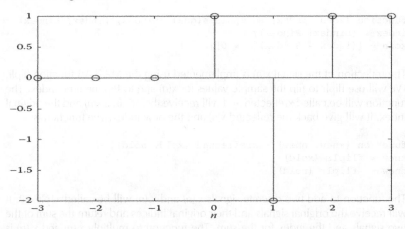

FIGURE 1.32   Signal for EOCE 1.7.

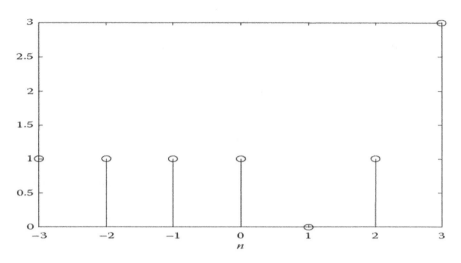

FIGURE 1.33   Signal for EOCE 1.7.

For the second signal, we have

$$3u(n-3) = \left\{ 0,0,0,0,0,0,3 \atop \uparrow \right\}$$

$$\delta(n-2) = \left\{ 0,0,0,0,0,1,0 \atop \uparrow \right\}$$

$$u(-n) = \left\{ 1,1,1,1,0,0,0 \atop \uparrow \right\}$$

and the sum is

$$x(n) = \left\{ 1,1,1,1,0,1,3 \atop \uparrow \right\}$$

We can use MATLAB to find this result using the script EOCE1_72.

The result is shown in Figure 1.33. Note in the previous script that the function xreflected was called and one of the passed parameters is stepsignal (0, −3, 3), which is a function that will return the signal $u(n)$. Thus, $u(n)$ and n are passed to xreflected.

EOCE 1.8

Write a general-purpose script to find the odd and even parts of a discrete signal $x(n)$ defined on the interval $n_1 \le n \le n_2$.

SOLUTION

Let us repeat the equations for the even and the odd part of $x(n)$.

$$x_{even} = (n) = \frac{1}{2}\left[x(n) + x(-n)\right]$$

$$x_{odd} = (n) = \frac{1}{2}\left[x(n) - x(-n)\right]$$

We can utilize the functions we have written so far to come up with the following script. Given $x(n)$ and its index n, we write the script EOCE1_8.

EOCE 1.9

Consider the following signal:

$$x(n) = \delta(n+3) + \delta(n+2) + \delta(n+1) - \delta(n-1) - \delta(n-2) - \delta(n-3)$$

for $-5 \le n \le 5$. Find the even and odd parts of $x(n)$.

SOLUTION

Although $x(n)$ can easily be seen as an odd signal, we will use MATLAB with the help of the functions and the scripts that we have written so far to plot the original signal, the reflected signal, the odd signal, and the even signal.

First we call the function impulsesignal six times to get $x(n)$ and then use the script EOCE1_9 for the odd and even parts of $x(n)$ to produce the plots.

The plots are shown in Figure 1.34.

EOCE 1.10

Consider the following signal:

$$x(n) = u(n-1) + \delta(n+1) \quad -2 \le n \le 2$$

Find and plot the following signals:

1. $x(-n)$
2. $x(n-2)$
3. $x(n) + x(-n)$

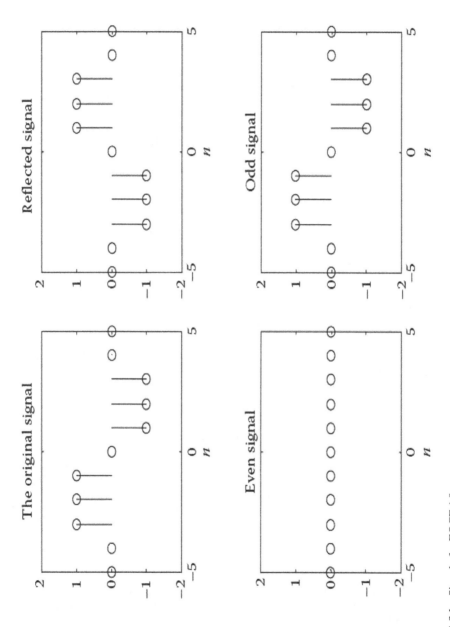

FIGURE 1.34 Signals for EOCE 1.9.

SOLUTION

1. Analytically,

$$x(n) = u(n-1) + \delta(n+1)$$

$$= \left\{0,0,0,\underset{\uparrow}{1},1\right\} + \left\{0,1,\underset{\uparrow}{0},0,0\right\} = \left\{0,1,\underset{\uparrow}{0},1,1\right\}$$

and $x(-n)$ is $\left\{1,1,\underset{\uparrow}{0},1,0\right\}$. Using MATLAB, we first generate $x(n)$ and then find its reflection using EOCE1_101 and the plot is shown in Figure 1.35.

2. Analytically,

$$x(n-2) = u(n-2-1) + \delta(n-2+1)$$

or we notice that $x(n-2)$ is $x(n)$ shifted by 2 and is

$$x(n-2) = \left\{0,1,\underset{\uparrow}{0},1,1\right\}$$

Using MATLAB, we use the shifting function we derived earlier to write the script EOCE1_102 and the plot is shown in Figure 1.36.

3. Using MATLAB, we can find $x(n) + x(-n)$ by writing the MATLAB script EOCE1_103 and the plot is shown in Figure 1.37.

EOCE 1.11

Audio signals are types of signals that are often used in signal processing and communications. In this exercise we will look at how can we use MATLAB, the sound card on your computer, and MATLAB data acquisition toolbox to acquire such a signal that you will speak out.

SOLUTION

To complete this task you will need a PC microphone or your laptop built-in microphone and a sound card that is also found on your PC or laptop. The sound card will work as the analogue-to-digital converter.

Let us start by acquiring 3 s of audio using the sound card. What we speak is an analogue signal and so the sound card will convert the speech to digital samples. Let us use 8000 Hz as the sampling rate on the sound card.

We will first create an analogue input object for the sound card. At the MATLAB prompt, we write

```
AI = analoginput ('winsound');
```

to create an input communication channel with the sound card. Then you need to add an input channel to the input object AI. This is done with the MATLAB command

```
addchannel (AI, 1);
```

to add a single channel to the input object.

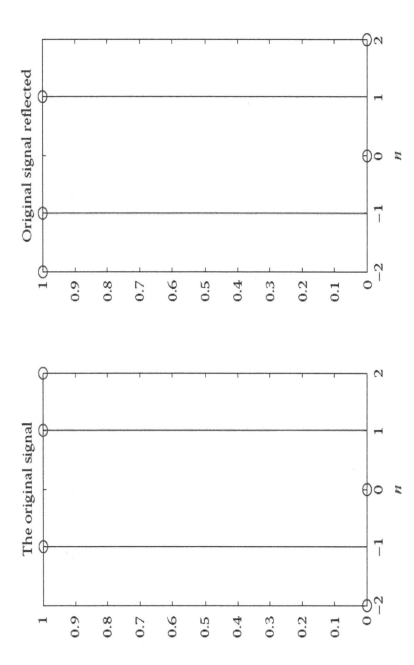

FIGURE 1.35 Signal for EOCE 1.10.

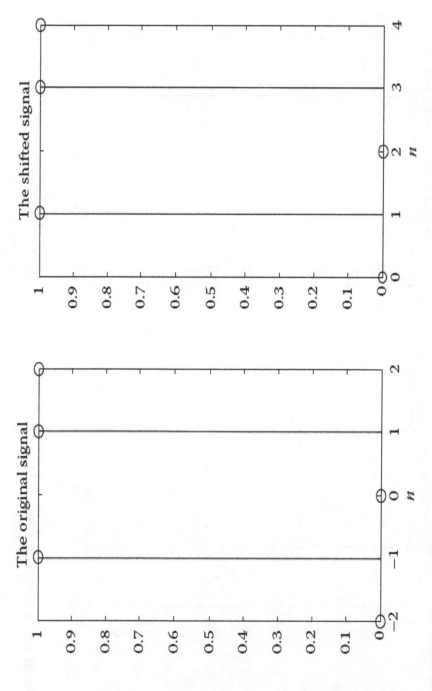

FIGURE 1.36   Signal for EOCE 1.10.

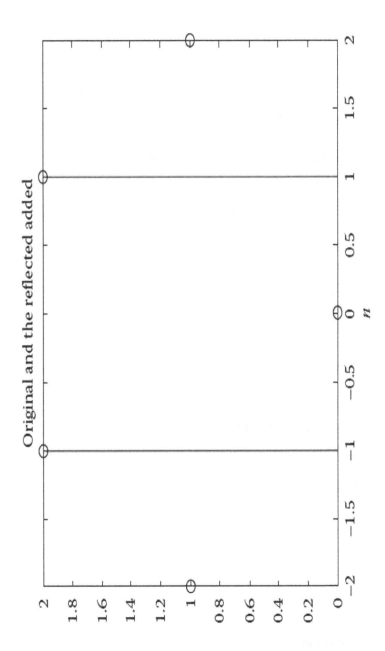

FIGURE 1.37 Signal for EOCE 1.10.

We need to specify a sampling rate (this topic will be revisited again in Chapter 8 for complete discussion). We will use 8000 Hz for that. We will write

```
Fs = 8000;
set (AI, 'SampleRate', Fs);
```

to define the sample rate to MATLAB and associate that with the input object AI.

We need to speak through the microphone for a limited time of 3 s and tell the channel how many samples per trigger we wish to collect. We do that with

```
Trig_duration = 3;
set (AI, 'SamplesPerTrigger', Trig_duration*Fs);
```

At this stage, we can start collecting data for the audio signal we will speak using the microphone. Before you hit enter after you type the command as follows, start speaking for about 5 s.

```
start (AI);
```

To collect the data samples of the words you just spoke, use the following command:

```
data = gatdata (AI);
```

It is always a good practice to delete the AI object after we finish. Write the command as follows:

```
delete (AI);
to delete the AI object.
```

At this time, you can plot the data you just collected.

1.23   END-OF-CHAPTER PROBLEMS

EOCP 1.1

Analytically find the following signals if $x(n) = nu(n-1) -\infty < n < \infty$

1. $x(-n)$
2. $x(-n + 1)$
3. $x(n) + x(-n)$
4. $x(2n)$
5. $x\left(\dfrac{n}{3}\right) + x(-n)$
6. $x(n)\, \delta\,(n-1)$
7. $x(-n)\, u(n-2) + \delta(n)$
8. $x(n-2) + \delta(n)\, x(n)$
9. $u\left(\dfrac{n}{2}\right) - x(n)$
10. $x(-n-2) + u(n-2)$

EOCP 1.2

Use MATLAB to generate the following signals if $x(n) = u(n) - u(n - 1)$ for $0 \le n \le 5$:

1. $x(-n)$
2. $x(n + 2)$
3. $x(n) + x(-n)$
4. $x(n - 2) + x(n + 2)$
5. $x(-n - 1)x(n)$
6. $x(-n) x(n) + x(-n - 1)$
7. $x(n) + \cos(2\pi n + \pi)$
8. $x(-n)\cos\left(3\pi n + \dfrac{\pi}{2}\right)$
9. $(.1)^n x(n)\cos\left(3\pi n + \dfrac{\pi}{2}\right)$

EOCP 1.3

Check the periodicity for each of the following signals for $0 \le n \le \infty$. If they are periodic, what is the period?

1. $\cos(2\pi n + \pi)$
2. $(.1)^n \cos\left(5\pi n + \dfrac{\pi}{2}\right)$
3. $u(n)$
4. $u(n) + 1$
5. $\cos\left(\sqrt{2}\pi n\right)$
6. $u(n) + \cos(2\pi n + \pi)$
7. $\cos(2\pi n + \pi) + \delta(n - 1)$
8. $2\cos(2n - \pi)$
9. $\cos\left(\dfrac{3}{2}n + \pi\right) + u(n)$

EOCP 1.4

Use MATLAB to check periodicity for the signal in EOCP 1.3.

EOCP 1.5

Find the power in the following signals:

1. $u(n)$ $n \ge 0$
2. $u(n)$ $n \ge 1$
3. $\displaystyle\sum_{m=0}^{\infty} \delta(n - m)$ $n \ge 0$

EOCP 1.6

Find the energy in each of the following signals for $-5 \leq n \leq 5$:

1. $u(n)$
2. $\cos(2\pi n)$
3. $nu(n)$
4. $2u(n)\cos(2\pi n)$
5. $u(n)\, u(-n)$
6. $n\cos(2\pi n)$

Find the energy in the following signals for $n > 0$:

1. $u(n)\,(.1)^n$
2. $(.1)^n \cos(2\pi n)$
3. $(.5)^n\, n$

EOCP 1.7

Consider the following signals:

1. $x(n) = u(n) + u(n-1)\ 0 \leq n \leq 5$
2. $x(n) = nu(n)\ 0 \leq n \leq 5$
3. $x(n) = (.1)^n \cos(2\pi n + 1)\ 0 \leq n \leq 5$
   a. Use MATLAB to sketch the even and the odd parts.
   b. Show that the energy in $x(n)$ is the sum of the energy in its components, the even and the odd parts.
   c. Are the signals bounded?

EOCP 1.8

Usually the discrete signals we deal with in engineering, $x(n)$, are obtained by taking samples from continuous signals $x(t)$. Give five examples where discrete signals are naturally discrete.

EOCP 1.9

Consider the following signals:

1. $x(t) = e^{-3t}u(t)$
2. $x(t) = e^{-t}\cos(1000t)u(t)$
   a. Let us take samples from both signals every 2 s. Find $x(n)$ for both.
   b. What is the time constant for the first signal?
   c. If $0 \leq n \leq 10$, find the energy in $x(n)$ for both signals.

EOCP 1.10

Let $y(n) = y(n-1) + u(n)$ with $y(-1) = 1$ for $n \geq 0$.

1. Write down the samples for $y(n)$.
2. Can you find a closed form equation for $y(n)$?

EOCP 1.11

Let $y(-1) = 1$ and consider the following equation:

$$y(n) = 2y(n-1) + u(n)$$

1. Find the samples for $y(n)$ for $n \geq 0$.
2. Find a mathematical closed form expression for $y(n)$.

2 Discrete System

2.1 DEFINITION OF A SYSTEM

A system is an assemblage of things that are combined to form a complex whole. When we study systems, we study the interactions and behaviors of such an assemblage when subjected to certain conditions. These conditions are called inputs. In its simplest case, a system can have one input and one output. This book deals with linear systems. We will call the input $x(n)$ and the output $y(n)$ as depicted in Figure 2.1.

2.2 INPUT AND OUTPUT

The discrete signal $x(n)$ is the continuous signal $x(t)$ sampled at $t = nT_s$, where t is the continuous time, n is an integer, and T_s is the sampling interval. If an input signal $x(t)$ is available at the input of a linear system, the system will operate on the signal $x(t)$ and produce an output signal that we call $y(t)$. As an example, consider the case of an elevator where you push a button to go to the fifth floor. Pushing the button is the input $x(t)$. The elevator is the system under consideration here. In addition to many other components, the elevator system consists of the small room to ride in and the motor that drives the elevator belt. The input signal $x(t)$ "asks" the elevator to move to the fifth floor. The elevator system will process this request and move to the fifth floor. The motion of the elevator to the selected floor is the output $y(t)$.

Pushing the button in this elevator case produces an electrical signal $x(t)$. This signal drives the motor of the elevator to produce a rotational motion, which is transferred, via some gears, to a translational motion. This translational motion is the output $y(t)$. To summarize, when an electrical input signal or request $x(t)$ is applied to the elevator system, the elevator will operate on the signal and produce $y(t)$, which, in this example, can be thought of as a translational motion.

The example that we just considered is inherently a continuous system. An example of a discrete or digital system is the digital computer. The computer has many inputs and outputs. An input can be generated by pressing any key on the keyboard and an output can be the display of the character that the user presses on the keyboard on the screen. The mouse and the scanner can be thought of as inputs as well. A computer program is a digital set of zeros and ones. It has an input set of raw data. The output of the computer program is the useful data.

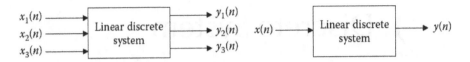

FIGURE 2.1 Linear system representation.

2.3 LINEAR DISCRETE SYSTEMS

A linear discrete system has the following properties:

1. If the input is $\alpha x_1(n)$, the output is $\alpha y_1(n)$.
2. If the input is $x_1(n) + x_2(n)$, the output is $y_1(n) + y_2(n)$.

By combining the two conditions, a system is considered linearly discrete if for the input $\alpha x_1(n) + \beta x_2(n)$, the output is $\alpha y_1(n) + \beta y_2(n)$, where α and β are constants.

Example 2.1

Consider the input–output relation.

$$y(n) = \left[\frac{R_2}{R_1 + R_2}\right] x(n)$$

where α and β are constants. Is this system linear?

SOLUTION

Using the definition of linearity introduced previously, we can proceed as follows. If the input is $x_1(n)$, the output is $y_1(n)$; therefore, we write

$$y_1(n) = \left[\frac{R_2}{R_1 + R_2}\right] x_1(n)$$

If the input is $x_2(n)$, the output is $y_2(n)$; therefore,

$$y_2(n) = \left[\frac{R_2}{R_1 + R_2}\right] x_2(n)$$

If the input is $\alpha x_1(n) + \beta x_2(n)$, the output is $y(n)$; therefore,

$$y(n) = \left[\frac{R_2}{R_1 + R_2}\right] \left(\alpha x_1(n) + \beta x_2(n)\right) = \alpha y_1(n) + \beta y_2(n)$$

Thus, the system is said to be linear.

Example 2.2

Consider the following system:

$$y(n) = \sqrt{x(n)}$$

Is this system linear?

SOLUTION

Consider two cases of the input signals, $x_1(n)$ and $x_2(n)$. The corresponding outputs are then given by

$$y_1(n) = \sqrt{x_1(n)}$$

for $x_1(n)$ and

$$y_2(n) = \sqrt{x_2(n)}$$

for $x_2(n)$. Now let us consider further that $\alpha x_1(n) + \beta x_2(n)$ has been applied to the system as its input. The corresponding output is then given by

$$y(n) = \sqrt{\alpha x_1(n) + \beta x_2(n)}$$

But

$$y(n) = \sqrt{\alpha x_1(n) + \beta x_2(n)} \neq \alpha y_1(n) + \beta y_2(n) = \alpha \sqrt{x_1(n)} + \beta \sqrt{x_2(n)}$$

Therefore, the system is nonlinear.

Example 2.3

Consider the following system:

$$y(n) = \left(\frac{2}{\left[2x(n) + 1 \right]} \right) x(n)$$

Is this system linear?

SOLUTION

As in Example 2.2, consider two cases of the input signals, $x_1(n)$ and $x_2(n)$. The corresponding outputs are then given by

$$y_1(n) = \left(\frac{2}{\left[2x_1(n) + 1 \right]} \right) x_1(n)$$

and

$$y_2(n) = \left(\frac{2}{\left[2x_2(n)+1\right]}\right)x_2(n)$$

Now, let us apply $\alpha x_1(n) + \beta x_2(n)$ to the system as its input. The corresponding output is then given by

$$y(n) = \left(\frac{2}{\left[2\left(\alpha x_1(n)+\beta x_2(n)\right)+1\right]}\right)\left(\alpha x_1(n)+\beta x_2(n)\right)$$

or

$$y(n) = \frac{\left[2\alpha x_1(n)+2\beta x_2(n)\right]}{\left[2\alpha x_1(n)+2\beta x_2(n)+1\right]}$$

$$\neq \alpha y_1(n) + \beta y_2(n)$$

$$= \frac{\left[2\alpha x_1(n)\right]}{2x_1(n)+1} + \frac{2\beta x_2(n)}{2x_2(n)+1}$$

Therefore, the system is nonlinear.

2.4   TIME INVARIANCE AND DISCRETE SIGNALS

A system is said to be time invariant if, for a shifted input $x(n - n_0)$, the output of the system is $y(n - n_0)$. To see if a system is time invariant or time variant, we do the following:

1. Find the output $y_1(n - n_0)$ that corresponds to the input $x_1(n)$.
2. Let $x_2(n) = x_1(n - n_0)$ and then find the corresponding output $y_2(n)$.
3. Check if $y_1(n - n_0) = y_2(n)$. If this is true then the system is time invariant. Otherwise it is time variant.

Example 2.4

Let $y(n) = \cos(x(n))$. Find out if the system is time variant or time invariant.

SOLUTION

STEP 1

$$y_1 = \cos\left(x_1(n)\right)$$

$y_1(n)$ shifted by n_0 is $y_1(n - n_0) = \cos\left(x_1(n - n_0)\right)$

STEP 2

$$y_2(n) = \cos\left(x_1(n - n_0)\right)$$

STEP 3

$$y_1(n - n_0) = y_2(n)$$

Thus, the system is time invariant.

Example 2.5

Let $y(n) = x(n)\cos(n)$. Find out if the system is time variant or time invariant.

SOLUTION

STEP 1

$$y_1(n) = x_1(n)\cos(n)$$

Therefore,

$$y_1(n - n_0) = x_1(n - n_0)\cos x_1(n - n_0)$$

STEP 2

$$y_2(n) = x_1(n - n_0)\cos(n)$$

STEP 3

$$y_1(n - n_0) \neq y_2(n)$$

Therefore, the system is time variant.

Example 2.6

Let $y(n) = ne^{-n}x(n)$. Find out if the system is time variant or time invariant.

SOLUTION

STEP 1

$$y_1(n) = ne^{-n}x_1(n)$$

Therefore,

$$y_1(n - n_0) = (n - n_0)e^{-(n - n_0)}x_1(n - n_0)$$

STEP 2

$$y_2(n) = ne^{-n}x_1(n - n_0)$$

STEP 3

$$y_1(n - n_0) \neq y_2(n)$$

Therefore, the system is time variant.

2.5   SYSTEMS WITH MEMORY

If at any value of n, $y(n)$ depends totally on $x(n)$ at that particular value, then in such a case, we say the system is without memory. Otherwise the system is with memory.

Example 2.7

Consider the input–output relation.

$$y_2(n) = (x(n))^2$$

Is the system with or without memory?

SOLUTION

For any value of n, $y(n)$ depends on $x(n)$ at that particular value. If we look at the output at $n = 4$, then we look at the input at $n = 4$ as well. In this case, the system is without memory.

Example 2.8

Consider the following system:

$$y(n) = nx(n - 1)$$

Is the system with or without memory?

SOLUTION

The output $y(n)$ at $n = 0$ depends on $x(n)$ at $n = -1$. Therefore, the system is with memory.

2.6   CAUSAL SYSTEMS

A causal system is a system where the output $y(n)$ at a certain time n_1 depends on the input $x(n)$ for $n < n_1$.

Example 2.9

If $x(n)$ is given as

$$x(n) = \left\{ \begin{matrix} 1, & 1, & 1 \\ \uparrow & & \end{matrix} \right\}$$

and the output $y(n)$ is

$$y(n) = \left\{ \begin{matrix} 1, & 1/2, & 1/4 \\ \uparrow & & \end{matrix} \right\}$$

Is this system causal?

SOLUTION

The first sample of the input has appeared at $n = 0$, as does the first sample of the output. Therefore, the system is causal. Note that the system is causal even if the output starts at any value for which $n \geq 0$.

Example 2.10

Let the input $x(n)$ be as

$$x(n) = \left\{ \begin{matrix} 1, & 0, & 1, & 0 \\ \uparrow & & & \end{matrix} \right\}$$

Let this signal be the input to a system where the output was recorded as

$$y(n) = \left\{ \begin{matrix} 1, & 1, & 0, & 1, & 0 \\ & & & \uparrow & \end{matrix} \right\}$$

Is the system causal?

SOLUTION

The first input sample starts at $n = 0$ and the first output sample starts at $n = -3$. This makes the system noncausal.

2.7 INVERSE OF A SYSTEM

If we can determine the input by measuring the output, then the system under consideration is said to be invertible. Note that if two inputs give the same output, then the system is not invertible.

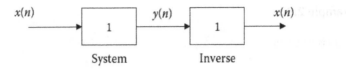

FIGURE 2.2   System for Example 2.11.

Example 2.11

Consider the following systems:

1. $y(n) = x(n)$
2. $y(n) = 2x(n)$
3. $y(n) = a\cos(x(n))$

Are these systems invertible?

SOLUTION

The first system is invertible. The corresponding pictorial representation is shown in Figure 2.2. The second system is also invertible. Figure 2.3 shows the corresponding pictorial depiction. The third system is not invertible. Why?

Let us consider two inputs, $x_1(n) = x(n)$ and $x_2(n) = x(n) + n\pi$, where n is an even integer. For this system, the output corresponding to $x_1(n) = x(n)$ is given by

$$y_1(n) = a\cos(x(n))$$

and the output corresponding to $x_2(n) = x(n) + n\pi$ is given by

$$y_2(n) = a\cos(x(n) + n\pi) = a\cos(x(n))$$

We can see here that two different inputs produced the same output. Therefore, the system is not invertible as claimed.

2.8   STABLE SYSTEM

The signal $x(n)$ is considered bounded if $|x(n)| < \beta < \infty$ for all n, where β is a real number. A system is said to be BIBO (bounded-input bounded-output) stable if and when the input is bounded, the output is also bounded. $y(n)$, the output, is bounded if $|y(n)| < \beta < \infty$.

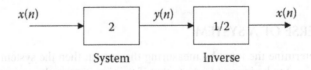

FIGURE 2.3   System for Example 2.11.

Example 2.12

Consider the system

$$y(n) = \sum_{k=0}^{M-1} x(n-k)$$

and assume that $x(n)$ is bounded. Is the system stable?

SOLUTION

If $x(n)$ is bounded, this implies that

$$|x(n)| < \beta$$

But a shifted version of $x(n)$ is also bounded for $x(n)$ is bounded. Therefore,

$$|y(n)| = \left| \sum_{k=0}^{M-1} x(n-k) \right| \leq M\beta$$

and the system is BIBO.

Example 2.13

Consider that $x(n)$ is bounded and applied to a system where $y(n)$ is obtained as

$$y(n) = n \sum_{k=0}^{M-1} x(n-k)$$

Is the system BIBO?

SOLUTION

Since $x(n)$ is bounded, we have $|x(n)| < \beta$. We also know that a shifted version of $x(n)$ is also bounded. Thus,

$$|y(n)| = \left| n \sum_{k=0}^{M-1} x(n-k) \right| < nM\beta$$

However, as n approaches infinity, $y(n)$ will grow without bounds and the system is not BIBO.

2.9   CONVOLUTION

To find the output of a discrete system $y(n)$ to an input $x(n)$, we need the impulse response $h(n)$ for the system. $h(n)$ is the output of the system when the input is $\delta(n)$, where $\delta(n)$ is the impulse signal. If we apply the signal $\delta(n)$ to the system as shown

FIGURE 2.4   The response to the impulse signal.

in Figure 2.4, the output is $h(n)$. Notice that we do not know the system itself (the input–output relation), but we know that if the input is the impulse signal $\delta(n)$, the output will be the impulse response $h(n)$.

Now consider the shifted impulse signal $\delta(n - m)$ to the same system, as shown in Figure 2.5. The output will also be shifted because we are considering only linear systems. Therefore, $\delta(n - m)$ will produce $h(n - m)$.

We have also seen in Chapter 1 that any discrete signal can be represented as the sum of weighted shifted impulses (samples). We have seen that the signal $x(n)$ can be represented as

$$x(n) = \sum_{m=-\infty}^{+\infty} x(m)\delta(n - m) \tag{2.1}$$

Note that each $x(m)$ is a sample and a constant. Also, if $x(m)$ is multiplied by $\delta(n - m)$ and applied to the discrete linear time-invariant system, the output will be $x(m)h(n - m)$, as shown in Figure 2.6.

Now let us say that we were to add all the shifted weighted samples

$$\sum_{m=-\infty}^{+\infty} x(m)\delta(n - m)$$

and present this as input to the same system. The output in this case is

$$\sum_{m=-\infty}^{+\infty} x(m)h(n - m)$$

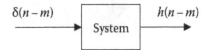

FIGURE 2.5   The response to the shifted impulse signal.

$$x(m)\delta(n - m) \longrightarrow \boxed{\text{System}} \longrightarrow x(m)h(n - m)$$

FIGURE 2.6   The response to the shifted impulse signal multiplied by a constant.

which is the sum of all the responses to each weighted sample individually. But

$$\sum_{m=-\infty}^{+\infty} x(m)\delta(n-m) = x(n)$$

Knowing that when $x(n)$ is presented as an input to a linear time-invariant system, the output is $y(n)$ allows us to write

$$y(n) = \sum_{m=-\infty}^{+\infty} x(m)h(n-m) \tag{2.2}$$

or

$$y(n) = x(n) * h(n)$$

The previous equation is the convolution equation that, given $x(n)$, the input to a discrete system, and $h(n)$, the impulse response, will give the output $y(n)$. This also tells that, given $h(n)$ for any system, you can find $y(n)$ for any input $x(n)$.

Example 2.14

Consider the system in which the impulse response is known to be

$$h(n) = (2)^n u(n) \quad n \geq 0$$

Find the output $y(n)$ for the input $x(n) = \delta(n)$.

SOLUTION

Note that we expect the output to be the $h(n)$ given because the input is $\delta(n)$. Using the convolution sum, we have

$$y(n) = \sum_{m=-\infty}^{\infty} x(m)h(n-m) = \sum_{m=-\infty}^{\infty} \delta(m)(2)^{n-m} u(n-m)$$

But $\delta(m)$ is only defined at $m = 0$. Therefore,

$$y(n) = \delta(0)(2)^{n-0} u(n-0) = 1(2)^n u(n) = h(n) \quad n \geq 0$$

as anticipated.

Example 2.15

Consider the input $x(n) = u(n)$ and the impulse response $h(n) = (.5)^n u(n)$ for a certain system. What is the output of the system?

SOLUTION

Using the convolution equation, we write

$$y(n) = \sum_{-\infty}^{+\infty} x(m)h(n-m) = \sum_{m=-\infty}^{+\infty} u(m)(.5)^{n-m}u(n-m)$$

But since $u(m) = 1$ for $n \geq 0$ and both $x(n)$ and $h(n)$ start at $n = 0$, we have

$$y(n) = \sum_{m=0}^{m=n}(.5)^{n-m} = (.5)^n \sum_{m=0}^{n}(.5)^{-m} = (.5)^n \sum_{m=0}^{n}\left((.5)^{-1}\right)^m$$

$$y(n) = (.5)^n \frac{\left[1-\left((.5)^{-1}\right)^{n+1}\right]}{1-(.5)^{-1}} \quad n \geq 0$$

The last result was obtained using the geometric series sum.

$$S = \sum_{m=0}^{m=n}(a)^m = \frac{\left[1-(a)^{n+1}\right]}{1-a}$$

Example 2.16

Let $x(n) = (.5)^n + (.6)^{n+1}$ and $h(n) = u(n)$ for $n \geq 0$. Find $y(n)$.

SOLUTION

Using the convolution equation, we write

$$y(n) = \sum_{m=-\infty}^{m=\infty} u(m)\left[(.5)^{n-m}+(.6)^{n-m+1}\right]u(n-m)$$

which reduces to

$$y(n) = \sum_{m=0}^{m=n}(.5)^{n-m}+(.6)^{n-m+1} = (.5)^n \frac{\left[1-\left((.5)^{-1}\right)^{n+1}\right]}{1-(.5)^{-1}} + (.6)^{n+1} \frac{\left[1-\left((.6)^{-1}\right)^{n+1}\right]}{1-(.6)^{-1}}$$

Example 2.17

Consider the system where $x(n) = Au(n)$ and $h(n) = Bu(n)$. Find the output $y(n)$.

SOLUTION

Using the convolution sum, we have

$$y(n) = \sum_{m=-\infty}^{+\infty} Au(m)Bu(n-m)$$

which reduces to

$$y(n) = \sum_{m=0}^{m=n} AB = AB(n+1) \; n \geq 0$$

Note that this system grows without bounds as n approaches infinity.

2.10 DIFFERENCE EQUATIONS OF PHYSICAL SYSTEMS

A difference equation that represents the input–output relation for a discrete linear time-invariant system is of the form as follows:

$$y(n) - \sum_{k=1}^{N} a_k y(n-k) = \sum_{k=0}^{M} b_k x(n-k) \tag{2.3}$$

with the initial conditions $y(-1)$, $y(-2)$, ... , $y(-N)$. The relation in the previous equation is a general relation between $x(n)$, the input, and $y(n)$, the output of the discrete system. The order of this general difference equation is N. We also say that N is the degree of the system. An example is the difference equation as follows:

$$y(n) - 2y(n-1) = x(n)$$

where N is 1, the order of the discrete difference equation; thus, it is a first-order difference equation. Our goal is to find the output $y(n)$ given the input $x(n)$.

The total solution has two parts: the homogeneous solution and the particular solution. We will learn how to find the total solution in the following section. But first, we will see how to find the homogeneous solution.

2.11 HOMOGENEOUS DIFFERENCE EQUATION AND ITS SOLUTION

The general form of a homogeneous difference equation is

$$y(n) - \sum_{k=1}^{N} a_k y(n-k) = 0 \tag{2.4}$$

with initial conditions $y(-1)$, $y(-2)$, ... , $y(-N)$ and all inputs set to zero. You can see that without inputs, the homogeneous solution is zero unless we have nonzero initial conditions.

To find $y_h(n)$, the homogeneous solution, we assume a solution of the form $c(p)^n$. If $c(p)^n$ is a solution, it must satisfy the homogeneous previous equation. So we substitute $c(p)^n$ into Equation (2.4) to get

$$c(p)^n - \sum_{k=1}^{N} a_k c(p)^{n-k} = 0$$

Since the summation in the previous equation is over k, we write

$$c(p)^n - c(p)^n \sum_{k=1}^{N} a_k p^{-k} = 0$$

By expanding the previous equation, we get

$$c(p)^n \left[1 - a_1 p^{-1} - a_2 p^{-2} - a_3 p^{-3} - \cdots a_N p^{-N} \right] = 0$$

To satisfy the previous equation, either

$$c(p)^n = 0$$

or

$$\left[1 - a_1 p^{-1} - a_2 p^{-2} - a_3 p^{-3} - \cdots - a_N p^{-N} \right] = 0$$

But $c(p)^n$ cannot be zero. Therefore,

$$\left[1 - a_1 p^{-1} - a_2 p^{-2} - a_3 p^{-3} - \cdots - a_N p^{-N} \right] = 0 \qquad (2.5)$$

This equation is called the characteristic equation of the system. If we multiply this equation by p^N, we will have

$$\left[p^N - a_1 p^{N-1} - a_2 p^{N-2} - a_3 p^{N-3} - \cdots - a_N \right] = 0$$

So the homogeneous solution $y_h(n)$ is

$$y_h(n) = c_1 (p_1)^n + c_2 (p_2)^n + \cdots + c_N (p_N)^n \qquad (2.6)$$

where $p_1, p_2, \ldots, p_N$ are the roots of

$$\left[p^N - a_1 p^{N-1} - a_2 p^{N-2} - a_3 p^{N-3} - \cdots a_N \right] = 0$$

and $c_1, c_2, \ldots, c_N$ are constant to be determined using the given initial conditions.

Example 2.18

Consider the homogeneous difference equation that describes a discrete system as

$$y(n) - 2y(n-1) = 0$$

with the initial condition $y(-1) = +1$. What is $y(n)$? Check your answer using iterations.

SOLUTION

The solution y(n) is the sum of the homogeneous and the particular parts. The homogeneous part is due to the initial condition and the particular part is due to the existing external inputs. In this case, there are no external inputs and so the particular part is zero. The homogeneous part is calculated by first assuming a solution of the form $y_h(n) = cp^n$ and then substituting in the given equation to get

$$cp^n - 2cp^{n-1} = 0$$

or

$$cp^n\left(1 - 2p^{-1}\right) = 0$$

The characteristic equation is then $1 - 2p^{-1} = 0$. Solving for p we get p = 2. The homogeneous solution is then

$$y_h(n) = c_1(2)^n$$

To find c_1, we use the initial condition and write

$$y(-1) = 1 = c_1(2)^{-1}$$

The final solution is then

$$y(n) = y_h(n) + y_p(n) = 2(2)^n u(n)$$

We can check the validity of the solution obtained by iteration. We can rewrite the given difference equation as

$$y(n) = 2y(n-1)$$

With the given initial condition, we can find y(n) for n ≥ 0 as in the following:

For n = 0, $y(0) = 2y(-1) = 2(1) = 2$
For n = 1, $y(1) = 2y(0) = 2(2) = 4$
For n = 2, $y(2) = 2y(1) = 2(4) = 8$
For n = 3, $y(3) = 2y(2) = 2(8) = 16$

These values can be checked via the closed form solution.

$$y(n) = 2(2)^n$$

For n = 0, $y(0) = 2(2)^0 = 2$
For n = 1, $y(1) = 2(2)^1 = 4$
For n = 2, $y(2) = 2(2)^2 = 8$
For n = 3, $y(3) = 2(2)^3 = 16$

If we are interested in the first few values of y(n), we can use iteration, and if we are interested, for example, in y(1000), we better find the closed form solution.

2.11.1 Case When Roots Are All Distinct

When all roots are distinct, the form of the homogeneous solution is

$$y_h(n) = c_1 p_1^n + c_2 p_2^n + \cdots + c_n p_n^n \tag{2.7}$$

where

$p_1, p_2, \ldots, p_n$ are the roots of the characteristic equation.
$c_1, c_2, \ldots, c_n$ are to be determined using the given initial conditions.

2.11.2 Case When Two Roots Are Real and Equal

If we consider a second-order discrete system, the roots in this case will be equal and real. Denoting the roots as $p_1 = p_2 = p$, the homogeneous solution in this case is

$$y_h(n) = c_1 p^n + c_2 n p^n \tag{2.8}$$

The reason for multiplying the second term by n is to make the two terms independent. If we have three equal real and repeating roots, the homogeneous solution is

$$y_h(n) = c_1 p^n + c_2 n p^n + c_3 n^2 p^n \tag{2.9}$$

If two roots are real and equal and one root is real, then

$$y_h(n) = c_1 p^n + c_2 n p^n + c_3 (p_3)^n \tag{2.10}$$

where $p = p_1 = p_2$ and p_3 is the other real root.

2.11.3 Case When Two Roots Are Complex

Suppose the roots in this case are p_1 and p_2. Complex roots always appear as complex conjugates. So if $p_1 = a + jb$, then $p_2 = a - jb$. Then we can put p_1 and p_2 in polar form and get

$$p_1 = Me^{j\theta}$$

$$p_2 = Me^{-j\theta}$$

and the homogeneous solution is then

$$y_h(n) = c_1 \left(Me^{j\theta}\right)^n + c_2 \left(Me^{-j\theta}\right)^n$$

For this solution to be real, c_1 must be the complex conjugate of c_2. Because of that we write the solution as

$$y_h(n) = c_1 \left(Me^{j\theta} \right)^n + c_1^* \left(Me^{-j\theta} \right)^n$$

Since c_1 is the complex conjugate of c_2, the two terms in the previous solution are conjugates. If

$$z_1 = a + jb$$

$$z_2 = a - jb$$

then

$$z_1 + z_2 = 2a = 2\text{real}(z_1) = 2\text{real}(z_2)$$

where real stands for the real part of the complex number. The constants can be written in polar form as $c_1 = Qe^{j\beta}$ and $c_2 = Qe^{-j\beta}$. With this at hand, we rewrite the homogeneous solution as

$$y_h(n) = Qe^{j\beta} \left(Me^{j\theta} \right)^n + Qe^{-j\beta} \left(Me^{-j\theta} \right)^n = 2\text{real}\left[Qe^{j\beta} \left(Me^{j\theta} \right)^n \right]$$

$$= 2\text{real}\left[QM^n e^{j(\beta + \theta n)} \right]$$

$$y_h(n) = 2QM^n \left(\text{real}\left[\cos(\theta n + \beta) + j\sin(\theta n + \beta) \right] \right)$$

$$= 2QM^n \cos(\theta n + \beta) \tag{2.11}$$

where

$\theta = \tan^{-1}(b/a)$
Q is the magnitude of c_1
$\beta = <c_1$ is the angle of the complex number c_1

The only two constants to be found in Equation (2.11) are β and Q. They can be found using the initial conditions.

2.12 NONHOMOGENEOUS DIFFERENCE EQUATIONS AND THEIR SOLUTIONS

The solution of the nonhomogeneous difference equation

$$y(n) - \sum_{k=1}^{N} a_k y(n-k) = \sum_{k=0}^{M} b_x x(n-k)$$

can be obtained by finding the homogeneous solution of

$$y(n) - \sum_{k=1}^{N} a_k y(n-k) = 0$$

with the initial conditions

$$y(-1), y(-2), ..., y(-N)$$

then adding it to the particular solution of

$$y(n) - \sum_{k=1}^{N} a_k y(n-k) = \sum_{k=0}^{M} b_x x(n-k)$$

So we write the total solution as

$$y(n) = y_h(n) + y_p(n)$$

Note that if we are given a nonhomogeneous difference equation with initial conditions, we first find $y_h(n)$ and do not evaluate the constants associated with $y_h(n)$ because these initial conditions are given for the total solution $y(n)$. The following example will illustrate this point.

Example 2.19

Consider the system described by the difference equation

$$y(n) - 2y(n-1) = x(n)$$

where $x(n) = u(n)$ for $n \geq 0$ and the initial conditions are $y(-1) = 1$.

SOLUTION

We start by finding the solution to the homogeneous part

$$y(n) - 2y(n-1) = 0$$

The characteristic equation is obtained by inspection as $p - 2 = 0$, and this gives $p = 2$. The homogeneous solution is then

$$y_h(n) = c_1(2)^n$$

We will not use the initial conditions to find c_1, but we will wait until we find $y_p(n)$ and then apply them to the total solution $y(n)$.

For the particular solution, and since our input is $u(n)$, a constant for $n \geq 0$, we predict that $y_p(n)$ is the constant k_1. This k_1 should be evaluated by substituting $y_p(n) = k_1$ in the given difference equation to obtain

$$k_1 - 2k_1 = 1 \quad \text{or} \quad k_1 = -1$$

Therefore, the total solution $y(n)$ is

$$y(n) = y_h(n) + y_p(n) = c_1 2^n - 1$$

In this total solution, we use the initial conditions to find c_1. We have $y(-1) = 1$. Thus,

$$y(-1) = c_1 2^{-1} - 1 = 1$$

and the constant is $c_1 = 4$. The total solution is then

$$y(n) = (4) 2^n - 1 \text{ for } n \geq 0$$

To test the validity of this solution, let us try to find the first few values of $y(n)$ by iteration. We were given that

$$y(n) - 2y(n-1) = u(n)$$

So

$$y(0) = 2y(-1) + u(0) = 2(1) + 1 = 3$$

and

$$y(1) = 2y(0) + u(1) = 2(3) + 1 = 7$$

In the derived closed form solution, $y(n) = (4)2^n - 1$, so $y(0) = (4)2^0 - 1 = 3$, and $y(1) = (4)2^1 - 1 = 7$.

2.12.1   How Do We Find the Particular Solution?

Usually the particular solution is the solution because of the inputs applied to the systems. There are no general rules as to the form of the particular solution. All we can do is guess. For a long time, certain particular solution forms were proven true for certain inputs. The forms are listed in Table 2.1 along with the given input $x(n)$.

TABLE 2.1

Particular Solutions for Selected Inputs

$x(n)$	$y(n)$
$ku(n)$	k_1
$k\alpha^n$	$k_1\alpha^n$
kn	$k_1 n + k_2$
$k\delta(n)$	0
$k\cos(n\theta)$	$k_1\cos(n\theta) + k_2\sin(n\theta)$
$k\sin(n\theta)$	$k_1\cos(n\theta) + k_2\sin(n\theta)$
$k\alpha^n\cos(n\theta)$	$k_1 a^n \cos(n\theta) + k_2 a^n \sin(n\theta)$

2.13   STABILITY OF LINEAR DISCRETE SYSTEMS: THE CHARACTERISTIC EQUATION

2.13.1   STABILITY DEPENDING ON THE VALUES OF THE POLES

We can determine if the discrete linear system is stable or not by evaluating the roots of the characteristic equation

$$p^N + a_1 p^{N-1} + a_2 p^{N-2} + \cdots + a_N = 0$$

If the roots are within the unit circle, then the system is stable. If any root is outside this range, then the system is unstable. In this case, we can solve for the eigenvalues or the poles of the system using the MATLAB® function roots. If the characteristic equation is given as

$$P^2 + 0.5p + 1 = 0$$

then using MATLAB, we type the command

```
roots([1 0.5 1])
```

to get the roots. But we are interested in the magnitude of the roots to make sure that this magnitude is within the unit circle. Thus, we modify the command and write

```
abs(roots([1 0.5 1]))
```

2.13.2   STABILITY FROM THE JURY TEST

Sometimes the real coefficients, $a_1, a_2, \ldots, a_N$, in the characteristic equation are not constants; they are variables on which the stability of the system depends. In such a case, we can use the Jury test to find out about the stability of the system. For higher order systems, the Jury test is superior.

To understand this test, we will consider the following general characteristic polynomial

$$a_5 p^5 + a_4 p^4 + a_3 p^3 + a_2 p^2 + a_1 p^1 + a_0 = 0$$

with a_5 different from zero and positive.

Next we arrange the leading coefficients as in the following table:

$a_5\ a_4\ a_3\ a_2\ a_1\ a_0$	
$[a_0\ a_1\ a_2\ a_3\ a_4\ a_5]a_0/a_5$	Subtract
$b_4\ b_3\ b_2\ b_1\ b_0\ 0$	
$[b_0\ b_1\ b_2\ b_3\ b_4]\ b_0/b_4$	Subtract
$c_3\ c_2\ c_1\ c_0\ 0$	
$[c_0\ c_1\ c_2\ c_3]c_0/c_3$	Subtract
$d_2\ d_1\ d_0\ 0$	
$[d_0\ d_1\ d_2]d_0/d_2$	Subtract
$e_2\ e_0\ 0$	
$[e_0\ e_1]e_0/e_1$	Subtract
$f_0\ 0$	

The first row in the table contains the coefficients in the characteristic equation in descending powers of p. The second row is the first row reversed. We then multiply the second row by the last element in row 1 divided by the last element in row 2, then subtract [row 2]a_0/a_5 from row 1 to get the third row. The process continues in this fashion until we get to the last row with one element only, f_0 in this case.

For the system represented by this characteristic equation to be stable, the leading boxed coefficients must all be positive. Let us look at an example to illustrate the process. Consider the characteristic following equation:

$$3p^2 + 2kp + 1 = 0$$

The table is arranged as in the following:

$3\ 2k\ 1$	
$[1\ 2k\ 3]1/3$	Subtract
$3 - 1(1/3) = [8/3] \quad 2k - 2k(1/3) = 4k/3$	0
$[4k/3\ 8/3](4k/3)/(8/3)$	
$8/3 - 4k/3\left[(4k/3)/(8/3)\right] = \dfrac{8 - 2k^2}{3}$	0

Stability requires that the boxed elements in the table be positive. This requires that

$$\frac{8 - 2k^2}{3} > 0$$

or

$$8 - 2k^2 = (2 - k)(2 + k) > 0$$

Thus, for stability, the values for k should be limited to $-2 < k < 2$.

Example 2.20

Consider the following system:

$$y(n) - 2y(n-1) = x(n)$$

Is the system stable?

SOLUTION

The character equation is

$$p - 2 = 0$$

with $p = 2$ as the root. Since $p > 1$, the system is unstable. With the Jury test, we have

1 -2	
$[-2 \ 1](-2/1)$	Subtract
-3	0

-3 is not positive. Thus, the system is unstable.

Example 2.21

Consider the following system:

$$y(n) - 0.5y(n-1) = x(n)$$

Is the system stable?

SOLUTION

The character equation is

$$p - 0.5 = 0$$

with $p = 5$. Since $0 \le 5 \le 1$, the system is stable.

With the jury test, we have

1 −0.5	
[−0.5 1](−0.5/1)	Subtract
[3/4]	0

3/4 is positive and the system is stable.

2.14   BLOCK DIAGRAM REPRESENTATION OF LINEAR DISCRETE SYSTEMS

So far we have seen linear discrete systems represented as difference equations. These systems can also be represented as block diagrams. The main components for the block diagrams are given next.

2.14.1   DELAY ELEMENT

The delay element is shown in Figure 2.7. In this case,

$$y(n) = x(n-1)$$

where D in the diagram is the delay time. The delay element can be implemented physically as a shift register.

2.14.2   SUMMING/SUBTRACTING JUNCTION

This junction is shown in Figure 2.8. The output here is

$$y(n) = x_1(n) \pm x_2(n)$$

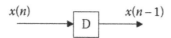

FIGURE 2.7   The delay element.

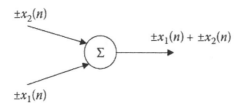

FIGURE 2.8   The summing/subtracting junction.

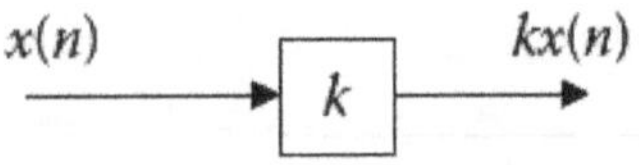

$$x(n) \xrightarrow{\hspace{2cm}} \boxed{k} \xrightarrow{kx(n)}$$

FIGURE 2.9 The multiplier element.

2.14.3 MULTIPLIER

The multiplier is shown in Figure 2.9. In this case,

$$y(n) = kx(n)$$

Example 2.22

Consider the following system:

$$y(n) - 2y(n-1) = x(n)$$

Represent this system in block diagram form.

SOLUTION

This is a first-order system where only $y(n)$ is delayed. The block is shown in Figure 2.10.

$$y(n) = 2y(n-1) + x(n) = 2Dy(n) + x(n)$$

The direction of the arrows is important. It means that the signals are flowing in the indicated direction.

Example 2.23

Consider the following system:

$$y(n) - 5y(n-1) - 3y(n-2) = x(n-1)$$

Give the block diagram representation.

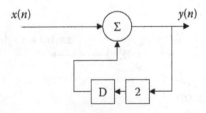

FIGURE 2.10 Block diagram for Example 2.22.

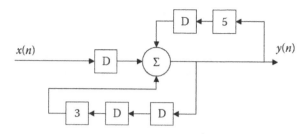

FIGURE 2.11 Block diagram for Example 2.23.

SOLUTION

We rewrite the output as

$$y(n) = 5y(n-1) + 3y(n-2) + x(n-1)$$

In terms of the D operator, we have

$$y(n) = 5Dy(n) + 3D^2y(n) + Dx(n)$$

The representation is shown in Figure 2.11.

2.15 FROM THE BLOCK DIAGRAM TO THE DIFFERENCE EQUATION

This is best illustrated by examples. The rule is that you find signals at the junctions proceeding from the input side going right to the output side.

Example 2.24

Consider the block diagram in Figure 2.12. Find the difference equation represented in the diagram.

SOLUTION

Looking at the output of the first summing junction from the left, we have the signal $x(n)$. Looking at the output of the second summing junction, we have the signal

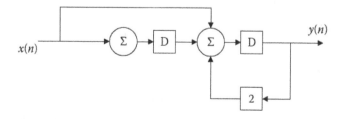

FIGURE 2.12 Block diagram for Example 2.24.

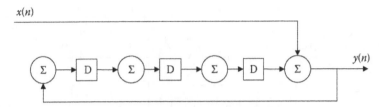

FIGURE 2.13   Block diagram for Example 2.25.

$x(n) + x(n-1) + 2y(n)$. The output of the last summing junction is $y(n)$. Coming to this last junction is the signal $x(n-1) + x(n-2) + 2y(n-1)$. Thus, the output of the last junction is $y(n) = 2y(n-1) + x(n-1) + x(n-2)$. The difference equation is then

$$y(n) - 2y(n-1) = x(n-1) + x(n-2)$$

Example 2.25

Consider the system shown in Figure 2.13. What is the difference equation?

SOLUTION

The output of the first summer is $y(n)$. The output of the second summer is $y(n-1)$. The output of the third summer is $y(n-2)$. The output of the fourth summer is $y(n)$, which is $y(n-3) + x(n)$. Therefore, the system representation as a difference equation is

$$y(n) - y(n-3) = x(n)$$

2.16   FROM THE DIFFERENCE EQUATION TO THE BLOCK DIAGRAM: A FORMAL PROCEDURE

This procedure will also be discussed using examples.

Example 2.26

Consider the following system:

$$y(n) - 2y(n-1) = x(n) + x(n-1)$$

Draw the block diagram.

SOLUTION

1. The given system is first order in $y(n)$. Therefore, we will need only one delay element.

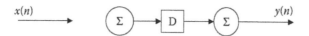

FIGURE 2.14 Block diagram for Example 2.26.

2. We initially draw the diagram shown in Figure 2.14 where we have one delay element preceded and followed by a summing junction. The input and output lines are drawn with $x(n)$ line not connected.
3. In the given equation, solve for $y(n)$ to get

$$y(n) = 2y(n-1) + x(n) + x(n-1)$$

Let us represent a delay by D, two delays by D^2 and so on to get

$$y(n) = 2Dy(n) + x(n) + Dx(n)$$

$$y(n) = D\left[2y(n) + x(n)\right] + x(n)$$

We will feed $(2y(n) + x(n))$ to the summer before the delay and $x(n)$ to the summer following the delay as shown in the final diagram in Figure 2.15.

Example 2.27

Consider the following system:

$$y(n) - 0.5y(n-1) - 0.3y(n-2) = 3x(n) + x(n-1)$$

Draw the block diagram.

SOLUTION

1. Our system is second order in $y(n)$, so we need two delay elements.
2. Draw the initial block diagram as shown in Figure 2.16 where every delay is preceded and followed by a summing junction and $x(n)$ is hanging and not connected.

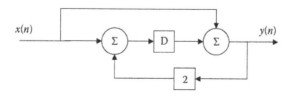

FIGURE 2.15 Block diagram for Example 2.26.

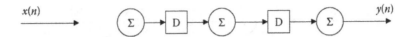

FIGURE 2.16 Block diagram for Example 2.27.

3. Solve for y(n) as

$$y(n) = 0.5y(n-1) + 0.3y(n-2) + 3x(n) + x(n-1)$$

Represent each delay by D and so on to get

$$y(n) = 0.5Dy(n) + 0.3D^2y(n) + 3x(n) + Dx(n)$$

$$y(n) = D^2[0.3y(n)] + D[0.5y(n) + x(n)] + 3x(n)$$

Feed 0.3y(n) to the summer before the first delay, (0.5y(n) + x(n)) to the summer before the second delay, and 3x(n) to the summer following the last delay to get the block diagram shown in Figure 2.17.

Example 2.28

Consider the following system:

$$y(n) - 3y(n-3) = x(n-3)$$

Draw the block diagram.

SOLUTION

1. The system is third order in y(n) and we will need three delay elements.
2. We initially draw the block as shown in Figure 2.18 with each delay preceded and followed by a summer and the x(n) input line hanging.

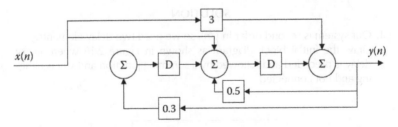

FIGURE 2.17 Block diagram for Example 2.27.

FIGURE 2.18 Block diagram for Example 2.28.

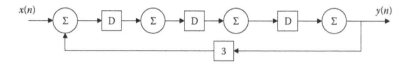

FIGURE 2.19   Block diagram for Example 2.28.

3. Next we solve for $y(n)$ as

$$y(n) = 3D^2 y(n) + D^2 x(n)$$

$$= \left[3y(n) + x(n) \right] D^3 + [0]D^2 + [0]D + [0]$$

$3y(n) + x(n)$ is fed to the summer preceding the first delay, 0 to the summer preceding the third delay, and 0 to the summer following the last delay as shown in Figure 2.19. Note that 0 means nothing is connected.

2.17   IMPULSE RESPONSE

The impulse response is the response due to an impulsive input. We will call the output $y(n)$, $h(n)$ when the input $x(n)$ is $\delta(n)$.

Example 2.29

Find the impulse responses for the following system:

$$y(n) - .5y(n-1) = x(n)$$

SOLUTION

The output will be $h(n)$ if $x(n) = \delta(n)$. The difference equation becomes

$$h(n) - .5h(n-1) = \delta(n)$$

This is a nonhomogeneous difference equation in $h(n)$ and has a solution that contains two parts: $h_h(n)$ and $h_p(n)$. For $h_h(n)$, the character equation is $p - .5 = 0$ and $p = 0.5$. So

$$h_h(n) = c_1 (.5)^n$$

For the particular solution we look at Table 2.1 and find that if the input is an impulse, the particular solution is zero. Thus, $h_p(n) = 0$. Therefore,

$$h(n) = h_h(n) + h_p(n) = c_1 (.5)^n$$

Next we find c_1. From the previous equation, we have

$$h(0) = c_1 (.5)^0 = c_1$$

But we do not know the values for $h(0)$. From the given system with $x(n) = \delta(n)$, we have

$$h(0) = .5h(-1) + \delta(0) = 0 + 1 = 1$$

Next we equate the $h(0)$ values to get

$$h(0) = 1 = h(0) = c_1$$

And finally

$$h(n) = (.5)^n \quad \text{for } n \geq 0$$

Example 2.30

Consider the following system:

$$y(n) + 5y(n-1) + 6y(n-2) = x(n)$$

What is the impulse response?

SOLUTION

The characteristic equation is

$$p^2 + 5p + 6 = 0$$

which gives $p_1 = -2$ and $p_2 = -3$.

Since $x(n) = \delta(n)$, the particular solution for $h(n)$ is $h_p(n) = 0$ as seen in Table 2.1. Thus, the total solution is

$$h(n) = c_1(-2)^n + c_2(-3)^n \quad n \geq 0$$

To find c_1 and c_2, we substitute $n = 0$ and $n = 1$ in the previous solution for $h(n)$ to get

$$h(0) = c_1 + c_2$$

$$h(1) = -2c_1 - 3c_2$$

But we do not know the values of $h(0)$ and $h(1)$. However, from the given system with $x(n) = \delta(n)$ and $n = 0$, we get

$$h(0) + 5h(-1) + 6h(-1) = \delta(0) = 1$$

with $h(0) = 1$.

And with $n = 1$, we have

$$h(1) + 5h(0) + 6h(-1) = 0$$

with $h(1) = -5$. Thus, we will end up with the two algebraic equations obtained by equating values for $h(0)$ and $h(1)$. The equations are

$$h(0) = 1 = c_1 + c_2$$

$$h(1) = -5 = -2c_1 - 3c_2$$

Solving the two equations gives $c_1 = -2$ and $c_2 = 3$ and the impulse response is then

$$h(n) = -2(-2)^n + 3(-3)^n \quad n \geq 0$$

2.18   CORRELATION

Correlation between two finite duration signals $x_1(n)$ and $x_2(n)$ is referred to as cross-correlation while correlation between the finite signal $x(n)$ and itself is referred to as auto-correlation. Next we explain both.

2.18.1   CROSS-CORRELATION

The cross-correlation between the two signals $x_1(n)$ and $x_2(n)$ is an indication of the similarities between the two signals as a function of the delay between them. The cross-correlation between $x_1(n)$ and $x_2(n)$ is written mathematically as

$$R_{x_1 x_2}(k) = \sum_{n=-\infty}^{\infty} x_1(n) x_2(n-k) \tag{2.12}$$

The index k is known as the lag of $x_1(n)$ relative to $x_2(n)$. If we let $n - k = m$, then the cross-correlation equation becomes

$$R_{x_1 x_2}(k) = \sum_{m=-\infty}^{\infty} x_1(m+k) x_2(m) = \sum_{n=-\infty}^{\infty} x_1(n+k) x_2(n) \tag{2.13}$$

This shows that we can use two equations to evaluate cross-correlation The question whether $R_{x_1 x_2}(k)$ is the same as $R_{x_1 x_2}(k)$ is examined next. According to the formula for cross-correlation, we have

$$R_{x_1 x_2}(k) = \sum_{n=-\infty}^{\infty} x_2(n) x_1(n-k)$$

But

$$R_{x_1 x_2}(k) = \sum_{n=-\infty}^{\infty} x_1(n+k) x_2(n)$$

and

$$R_{x_1 x_2}(-k) = \sum_{n=-\infty}^{\infty} x_1(n) x_2(n+k) = \sum_{n=-\infty}^{\infty} x_1(n-k) x_2(n)$$

Thus, we see that

$$R_{x_1 x_2}(k) = R_{x_1 x_2}(-k) \tag{2.14}$$

The previous equations for cross-correlation are defined for energy signals where the summation converges to some constant. If the signals are power signals (periodic signals are examples of this type), then the summation will not converge, and thus, we will use average values. In this case if the period of the discrete sequence is N then the cross-correlation is taken over one period (which is the same as averaging over an infinite interval) and is defined as

$$R_{x_1 x_2}(k) = \frac{1}{N} \sum_{n=0}^{N} x_1(n) x_2(n-k) \tag{2.15}$$

We have also seen before that the convolution between the two signals $x(n)$ and $h(n)$ is given by

$$y(k) = \sum_{n=-\infty}^{\infty} x(n) h(k-n)$$

and the convolution between $x(n)$ and $h(-n)$ is

$$y(k) = \sum_{n=-\infty}^{\infty} x(n) h(k+n) \tag{2.16}$$

But the quantity to the right of the equal sign in Equation (2.16) is nothing but the cross-correlation between $x(n)$ and $h(n)$. Therefore, we conclude that the convolution between $x(n)$ and $h(-n)$ is the cross-correlation between $x(n)$ and $h(n)$. We write

$$x(n) * h(-n) = R_{xh}(k) = \sum_{n=-\infty}^{\infty} x(n) h(k+n) \tag{2.17}$$

2.18.2 Auto-correlation

Auto-correlation is defined between the signal and itself. It is defined for energy signals as

$$R_{xx}(k) = \sum_{n=-\infty}^{\infty} x(n) x(n-k) \tag{2.18a}$$

and for power signals of period N it is defined as

$$R_{xx}(k) = \frac{1}{N} \sum_{n=0}^{N} x(n)x(n-k) \tag{2.18b}$$

2.19 SOME INSIGHTS

Let us say that we have a first-order system with the output $y(n)$ given as

$$y(n) = (0.5)^n \quad \text{for } n > 0$$

As n approaches infinity, the output will approach the value zero. In this sense we say the output is stable for our particular input. For first-order systems (that are described by first-order difference equations), the output will have one term of the form $(a)^n$. For second-order systems, the output will have two terms similar to $(a)^n$ at the most. For third-order systems, we will have three terms and so on.

In many systems of order greater than two, and for the purpose of analysis and design, we can reduce the order of the system at hand to a second-order system due to the fast decay of some of these terms. The solution for the output for these systems is in the following form:

$$y(n) = c_1 (a_1)^n + c_2 (a_2)^n$$

The stability of the system is determined by the values of a_1 and a_2. If any of the as has a magnitude that is greater than 1, the output $y(n)$ will grow wild as n approaches infinity. If all the as have a magnitude not greater than 1, then the output $y(n)$ will decay gradually and stays at a fixed value as n progresses The as are called the eigenvalues of the system. Therefore, we can say that a linear time-invariant system is stable if the eigenvalues of the system have magnitudes not greater than 1.

2.19.1 HOW CAN WE FIND THESE EIGENVALUES?

A linear time-invariant system can always be represented by a linear difference equation with constant coefficients.

$$y(n) - \sum_{k=1}^{N} a_k y(n-k) = \sum_{k=0}^{M} b_k x(n-k)$$

We can look at the auxiliary algebraic equation by setting the input $x(n)$ to zero,

$$y(n) - \sum_{k=1}^{N} a_k y(n-k) = 0$$

and letting $y(n - k) = D^k y(n)$ to get

$$y(n) - \sum_{k=1}^{N} a_k D^k y(n) = 0$$

We can expand the previous equation to get

$$y(n) - a_1 D y(n) - a_2 D^2 y(n) - \cdots - a_N D^N y(n) = 0$$

We can factor out $y(n)$ as

$$y(n)\left[1 - a_1 D - a_1 D^2 - \cdots - a_N D^N\right] = 0$$

$y(n)$ cannot be zero (in which case the output of the system would be zero at all times), and therefore,

$$\left[1 - a_1 D - a_2 D^2 - \cdots - a_N D^N\right] = 0$$

or

$$\left[\frac{1}{a_N} - \frac{a_1}{a_N} D - \frac{a_2}{a_N} D^2 - \cdots - D^N\right] = 0$$

This is an Nth order algebraic equation with N roots. Let us call them the N as. These are the eigenvalues of the system.

2.19.2 STABILITY AND EIGENVALUES

To summarize, any linear time-invariant system can be modeled by a linear difference equation with constant coefficients. The auxiliary algebraic equation that can be obtained from the difference equation will have a number of roots called the eigenvalues of the system. The stability of the system is determined by these roots. These roots may be real or complex. If *all* the magnitudes of the roots are less than or equal to 1, then the system is stable. If *any* of the roots has a magnitude that is greater than 1, the system is unstable. The eigenvalues of the system are responsible for the shape of the output $y(n)$. They dictate the shape of the transients of the system as well.

2.20 END-OF-CHAPTER EXAMPLES

EOCE 2.1

Are the following systems linear?

    1. $y(n) = (.5)^n x(n) + 1$

2. $y(n) = (.5)^n \cos(2x(n))$
3. $y(n) = \sin(n) - x(n)$

SOLUTION

1. For an input $x_1(n)$ the output is

$$y_1(n) = (.5)^n x_1(n) + 1$$

For an input $x_2(n)$ the output is

$$y_2(n) = (.5)^n x_2(n) + 1$$

If the input is $\alpha x_1(n) + \beta x_2(n)$, then the output is

$$y(n) = (.5)^n \left[\alpha x_1(n) + \beta x_2(n) \right] + 1$$

$$= (.5)^n \alpha x_1(n) + (.5)^n \beta x_2(n) + 1$$

But $\alpha y_1(n) + \beta y_2(n)$ is

$$\alpha (.5)^n x_1(n) + \beta (.5)^n x_2(n) + 2 \neq \alpha (.5)^n x_1(n) + \beta (.5)^n x_2(n) + 1$$

Thus, the system is not linear.
2. For $y = (.5)^n \cos(2x(n))$ and if the input is $x_1(n)$ then

$$y_1(n) = (.5)^n \cos(2x_1(n))$$

If the input is $x_2(n)$, then

$$y_2(n) = (.5)^n \cos(2x_2(n))$$

Now

$$\alpha y_1(n) + \beta y_2(n) = \alpha (.5)^n \cos(2x_1(n)) + \beta (.5)^n \cos(2x_2(n))$$

If the input is $\alpha x_1(n) + \beta x_2(n)$, then

$$y(n) = (.5)^n \cos(2\alpha x_1(n) + \beta x_2(n))$$

which is clearly not equal to $\alpha y_1(n) + \beta y_2(n)$. Thus, the system is not linear.
3. For $y = \sin(n) - x(n)$, if the input is $x_1(n)$, then

$$y_1(n) = \sin(n) - x_1(n)$$

If the input is $x_2(n)$, then

$$y_2(n) = \sin(n) - x_2(n)$$

Now

$$\alpha y_1(n) + \beta y(n) = \alpha\big(\sin(n) - x_1(n)\big) + \beta\big(\sin(n) - x_2(n)\big)$$

If the input is $\alpha x_1(n) + \beta x_2(n)$,

$$y(n) = \sin(n) - \big(\alpha x_1(n) + \beta x_2(n)\big)$$

$$= \sin(n) - \alpha x_1(n) - \beta x_2(n)$$

which is clearly not equal to $\alpha y_1(n) + \beta y_2(n)$. Thus, the system is not linear. We can use MATLAB to generate the sequence $\alpha(\sin(n) - x_1(n)) + \beta(\sin(n) - x_2(n))$ and the sequence $\sin(n) - \alpha x_1(n) - \beta x_2(n)$, and find the difference between them. If the difference is zero, then the system is linear. Otherwise it is nonlinear. We will let $\alpha = \beta = 1$ and $x_1(n) = n$ and $x_2(n) = 2n$. The MATLAB script is EOCE2_1.

The answer will be 1.4112. Because e in the previous script is not zero, this proves that the system is nonlinear.

EOCE 2.2

Consider the same systems as in EOCE 2.1. Are they time-variant systems?

SOLUTION

1. For $x_1(n)$ as an input,

$$y_1(n) = (.5)^n x_1(n) + 1$$

$y_1(n)$ shifted by n_0 is

$$y_1(n - n_0) = (.5)^{n-n_0} x_1(n - n_0) + 1$$

If we apply a shifted version of $x_1(n)$, $x_1(n - n_0)$, then

$$y_1(n) = (.5)^n x_1(n - n_0) + 1$$

But $y_2(n) \neq y_1(n - n_0)$, so the system is time variant.
2. For $x_1(n)$ as an input

$$y_1(n) = (.5)^n \cos\big(2x_1(n)\big)$$

$y_1(n)$ shifted by n_0 is

$$y_1(n - n_0) = (.5)^{n-n_0} \cos(2x_1(n - n_0))$$

If we apply $x_1(n - n_0)$ to the system, the output would be

$$y_2(n) = (.5)^n \cos(2x_1(n - n_0))$$

But $y_1(n - n_0) \neq y_2(n)$, so the system is time variant.

3. For $y = \sin(n) - x(n)$, if we apply $x_1(n)$ as input, the output is

$$y_1(n) = \sin(n) - x_1(n)$$

$y_1(n)$ shifted by n_0 is

$$y_1(n - n_0) = \sin(n - n_0) - x_1(n - n_0)$$

If we apply $x_1(n - n_0)$ to the system, then the output is

$$y_2(n) = \sin(n) - x_1(n - n_0)$$

But $y_1(n - n_0) \neq y_2(n)$, so it is a time-variant system.
We can use MATLAB to prove these results. We will compare $y_1(n - n_0)$ to $y_2(n)$ given earlier. Let $x(n) = n$, and let us shift by two samples. The script is EOCE2_2.
    The result will be 2.378. This proves the analytical result.

EOCE 2.3

Consider the following systems for $n \geq 0$:

1. $y(n) = (.5)^n x(n)$
2. $y(n) = n(.5)^n x(n)$
3. $y(n) = (.5)^n x(n)$

If $x(n)$ is bounded, what about $y(n)$?

SOLUTION

1. Since $x(n)$ is bounded, we write

$$|x(n)| < \beta < \infty$$

If $(.5)^n$ is bounded, then we will conclude that $y(n)$ is bounded too.

$$\sum_{n=0}^{N} (.5)^n = \frac{1 - (.5)^{N+1}}{1 - .5}$$

as $N \rightarrow \infty$, $(.5)^{N+1}$ goes to zero and

$$\sum_{n=0}^{\infty}(.5)^n = \frac{1}{1-.5} = 2$$

Thus, $y(n)$ is also bounded.

2. Since $x(n)$ is bounded, we will check the signal $n(.5)^n$. As n approaches ∞, $(.5)^n$ will approach zero. The multiplicative term $n(.5)^n$ will die as n approaches infinity, so the system output is bounded.

3. Because $x(n)$ is known to be bounded, we will check the sum $(5)^n$ for $n \geq 0$.

$$\sum_{n=0}^{N}(5)^n = \frac{1-(5)^{N+1}}{1-5}$$

As N approaches infinity, $(5)^{N+1}$ will grow without bounds. Thus,

$$\sum_{n=0}^{N}(5)^n$$

will not converge to any value. Hence, the system output is not bounded.

EOCE 2.4

Consider the following finite duration signals:

1. $x_1(n) = \left\{ \begin{matrix} 1, 1, 1 \\ \uparrow \end{matrix} \right\}$ $x_2(n) = \left\{ \begin{matrix} 1, 1, 1 \\ \uparrow \end{matrix} \right\}$

2. $x_1(n) = \left\{ \begin{matrix} 1, 1, 1 \\ \uparrow \end{matrix} \right\}$ $x_2(n) = \left\{ \begin{matrix} 1, 1, 1 \\ \quad\uparrow \end{matrix} \right\}$

3. $x_1(n) = \left\{ \begin{matrix} 1, 1, 1 \\ \uparrow \end{matrix} \right\}$ $x_2(n) = \delta(n) = \left\{ \begin{matrix} 1 \\ \uparrow \end{matrix} \right\}$

Find $x_1(n) * x_2(n)$, the convolution result, for the aforementioned cases.

SOLUTION

For two finite signals $x_1(n)$ and $x_2(n)$, and if $x_1(n)$ is defined on the interval $sn_1 < n_1 < en_1$ and $x_2(n)$ is defined on the interval $sn_2 < n_2 < en_2$, then $y(n) = x_1(n) * x_2(n)$ will start at the index $sn_1 + sn_2$ and ends at the index $en_1 + en_2$.

1. $y(n) = \sum_{0}^{2} x_1(m) x_2(n-m)$

In this case, $x_1(n)$ is defined on $0 < n_1 < 2$ and $x_2(n)$ is defined on the interval $0 < n_2 < 2$. $y(n)$ will start at $0 + 0 = 0$ and ends at $2 + 2 = 4$. So we write

$$y(0) = x_1(0)x_2(0) + x_1(1)x_2(-1) + x_1(2)x_2(-2) = 1 + 0 + 0 = 1$$

$$y(1) = x_1(0)x_2(1) + x_1(1)x_2(0) + x_1(2)x_2(-1) = 1 + 1 + 0 = 2$$

$$y(2) = x_1(0)x_2(2) + x_1(1)x_2(1) + x_1(2)x_2(0) = 1 + 1 + 1 = 3$$

$$y(3) = x_1(0)x_2(3) + x_1(1)x_2(2) + x_1(2)x_2(1) = 0 + 1 + 1 = 2$$

$$y(4) = x_1(0)x_2(4) + x_1(1)x_2(3) + x_1(2)x_2(2) = 0 + 0 + 1 = 1$$

Therefore, the convolution result is

$$y(n) = x_1(n) * x_2(n) = \left\{ \underset{\uparrow}{1}, 2, 3, 2, 1 \right\}$$

We can use MATLAB to find $y(n) = x_1(n) * x_2(n)$ if both signals start at $n = 0$. Since the signals here both start at $n = 0$, we can use MATLAB and write the script EOCE2_41 to get the same result as shown in Figure 2.20.

2. In this case, $x_1(n)$ starts at $n = 0$ and ends at $n = 2$, and $x_2(n)$ starts at $n = -2$ and ends at $n = 0$. $y(n) = x_1(n) * x_2(n)$ will start at $0 + (-2) = -2$ and ends at $n = 2 - 2 = 0$. So we write

$$y(n) = \sum_{m=0}^{2} x_1(m)x_2(n-m)$$

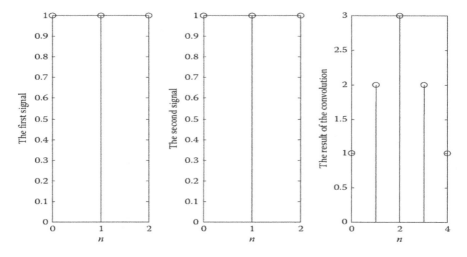

FIGURE 2.20 Signals for EOCE 2.4.

and

$$y(-2) = x_1(0)x_2(-2-0) + x_1(1)x_2(-2-1) + x_1(2)x_2(-2-2)$$
$$= (1)(1) + (1)(0) + (1)(0) = 1$$

$$y(-1) = x_1(0)x_2(-1-0) + x_1(1)x_2(-1-1) + x_1(2)x_2(-1-2)$$
$$= (1)(1) + (1)(1) + (1)(0) = 2$$

$$y(0) = x_1(0)x_2(0-0) + x_1(1)x_2(0-1) + x_1(2)x_2(0-2)$$
$$= (1)(1) + (1)(1) + (1)(1) = 3$$

$$y(1) = x_1(0)x_2(1-0) + x_1(1)x_2(1-1) + x_1(2)x_2(1-2)$$
$$= (1)(0) + (1)(1) + (1)(1) = 2$$

$$y(2) = x_1(0)x_2(2-0) + x_1(1)x_2(2-1) + x_1(2)x_2(2-2)$$
$$= (1)(0) + (1)(0) + (1)(1) = 1$$

Therefore, we have

$$y(n) = x_1(n) * x_2(n) = \left\{1, 2, \underset{\uparrow}{3}, 2, 1\right\}$$

We can use MATLAB to find $y(n)$, but, since $x_2(n)$ does not start at $n = 0$, we will do some modification before we can use the conv function from MATLAB. The MATLAB script EOCE2_42 will do that.

The plot is shown in Figure 2.21.

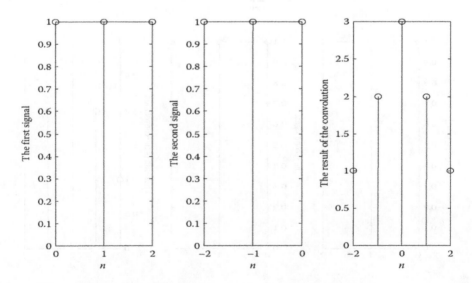

FIGURE 2.21 Signals for EOCE 2.4.

3. The starting index of $x_1(n)$ is $sn_1 = 0$ and the ending index is $en_1 = 2$. The starting index of $x_2(n)$ is $sn_2 = 0$ and the ending index is $en_2 = 0$. Therefore, $y(n)$ will start at $ns = sn_1 + sn_2 = 0 + 0 = 0$ and will end at $en = en_1 + en_2 = 2$.

$$y(n) = \sum_{m=0}^{2} x_1(m) x_2(n-m)$$

$$y(0) = x_1(0) x_2(0-0) + x_1(1) x_2(0-1) + x_1(2) x_2(0-2)$$
$$= (1)(1) + (1)(0) + (1)(0) = 1$$

$$y(1) = x_1(1) x_2(1-0) + x_1(1) x_2(1-1) + x_1(1) x_2(1-2)$$
$$= (1)(0) + (1)(1) + (1)(0) = 1$$

$$y(2) = x_1(2) x_2(2-0) + x_1(2) x_2(2-1) + x_1(2) x_2(2-2)$$
$$= (1)(0) + (1)(0) + (1)(1) = 1$$

$$y(n) = x_1(n) * x_2(n) = \left\{ \underset{\uparrow}{1}, 1, 1 \right\}$$

We can use MATLAB here directly since both signals start at $n = 0$ and write the script EOCE2_43.
The result is shown in Figure 2.22.

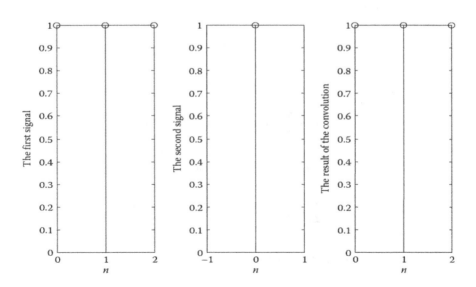

FIGURE 2.22 Signals for EOCE 2.4.

EOCE 2.5

Consider the following system represented by the impulse response $h(n)$

$$h(n) = (.5)^n u(n)$$

with

1. $x(n) = u(n)$
2. $x(n) = (.5)^n u(n)$

Find $y(n) = x(n) * h(n)$ for both cases.

SOLUTION

1. For the first case,

$$y(n) = \sum_{m=-\infty}^{+\infty} u(m)(.5)^{n-m} u(n-m)$$

Since both signals start at $n = 0$, $y(n)$ will start at $n = 0$.

$$y(n) = \sum_{m=0}^{m=n} (.5)^{n-m} = (.5)^n \sum_{m=0}^{m=n} (.5^{-1})^m = (.5)^n \left[\frac{1 - \left((.5)^{-1}\right)^{n+1}}{1 - (.5)^{-1}} \right] \quad n \geq 0$$

After simplification, we get

$$y(n) = 2 - (.5)^n \quad n \geq 0$$

2. For the second case,

$$y(n) = \sum_{m=-\infty}^{+\infty} u(m)(.5)^m (.5)^{n-m} u(n-m)$$

Since both $x(n)$ and $h(n)$ in this case start at $n = 0$, we have

$$y(n) = \sum_{m=0}^{m=n} (.5)^m (.5)^n (.5)^{-m} = \sum_{m=0}^{n} (.5)^n = (.5)^n \sum_{m=0}^{n} 1$$

And after simplifying, we arrive at

$$y(n) = n(.5)^n \quad n \geq 0$$

EOCE 2.6

Consider the following homogeneous difference equations:

1. $y(n) + .6y(n - 1) = 0$ with $y(-1) = 1$
2. $y(n) + y(n - 1) + (1/4) y(n - 2) = 0$ with $y(-1) = y(-2) = 0$
   a. Find the characteristic equation for the aforementioned systems and see if they are stable.
   b. Find the homogeneous solution for both. The homogeneous solution is the solution due to the initial conditions.

SOLUTION

For the first system, the characteristic equation is

$$p + .6 = 0 \quad \text{or} \quad p = -.6$$

Since $0 < .6 < 1$, the system is stable. The homogeneous solution is

$$y_h(n) = c_1(-.6)^n \quad \text{for } n \geq 0$$

With $y(-1) = 1$, we have

$$1 = c_1(-.6)^{-1}$$

$$c_1 = -.6$$

Thus, the solution is

$$y_h(n) = -0.6(-.6)^n \quad \text{for } n \geq 0$$

For the second system, the characteristic equation is

$$p^2 + p + \frac{1}{4} = 0$$

By factoring, we get

$$\left(p + \frac{1}{2}\right)\left(p + \frac{1}{2}\right) = 0$$

$$p_1 = p_2 = -\frac{1}{2}$$

Since $0 < .5 < 1$, the system is stable. The homogeneous solution is then

$$y_h = c_1\left(-\frac{1}{2}\right)^n + nc_2\left(-\frac{1}{2}\right)^n$$

With the initial conditions given, we can write

$$y(-1) = 1 = -2c_1 + 2c_2$$

$$y(-2) = 0 = 4c_1 - 8c_2$$

Solving these equations, we arrive at $c_2 = -(1/2)$ and $c_1 = -1$. The final solution is then

$$y_n(n) = -\left(-\frac{1}{2}\right)^n - \frac{1}{2}\left(-\frac{1}{2}\right)^n n \quad n \geq 0$$

EOCE 2.7

Consider the following difference equations:

1. $y(n) + .6y(n-1) = u(n)$, $y(-1) = 1$
2. $y(n) + y(n-1) + \frac{1}{4}y(n-2) = \delta(n), y(-1) = 1; y(-2) = 0$ 2

Find the total solutions for the two systems.

SOLUTION

For the first system, the characteristic equation is $p + .6 = 0$ or $p = -.6$ and the homogeneous part of the solution is

$$y_h(n) = c_1(-.6)^n$$

The particular solution is taken to be

$$y_p(n) = c_2$$

This particular solution is taken from Table 2.1 with $x(n) = u(n)$, a constant. We will substitute the particular solution into the difference equation given. We will get

$$c_2 + .6c_2 = 1$$

and this gives us $c_2 = 1/1.6$. The total solution now is

$$y(n) = y_h(n) + y_p(n) = c_1(-.6)^n + \frac{1}{1.6}$$

To find c_1, we use the initial condition given to us and write

$$y(-1) = 1 = c_1\left(\frac{1}{-.6}\right) + \frac{1}{1.6}$$

$$c_1 = \frac{6}{16} - .6$$

and finally the solution is

$$y(n) = \left(\frac{6}{16} - .6\right)(-.6)^n + \frac{1}{1.6} \quad n \geq 0$$

For the second system, the characteristic equation is

$$p^2 + p + \frac{1}{4} = 0 \quad \text{with} \quad p_1 = p_2 = -\frac{1}{2}$$

The homogeneous solution is

$$y_h(n) = c_1\left(-\frac{1}{2}\right)^n + c_2 n\left(-\frac{1}{2}\right)^n$$

The particular solution is $y_p(n) = 0$ and is obtained from Table 2.1. The total solution is

$$y(n) = c_1\left(-\frac{1}{2}\right)^n + nc^2\left(-\frac{1}{2}\right)^n \quad n \geq 0$$

Next we find the constants c_1 and c_2 using the given initial conditions.

$$y(-1) = +1 = -2c_1 + 2c_2$$

$$y(-2) = 0 = 4c_1 - 8c_2$$

Solving the previous equation gives $c_1 = -1$ and $c_2 = -1/2$, and the total solution is then

$$y(n) = -\left(-\frac{1}{2}\right)^n - \frac{1}{2}n\left(-\frac{1}{2}\right)^n \quad n \geq 0$$

EOCE 2.8

Consider the following system:

$$y(n) + y(n-1) = x(n)$$

For $x(n) = u(n)$, find $y(n)$ due to the input $x(n)$ using the convolution summation equation.

SOLUTION

We will first find $h(n)$ and then find $y(n)$ using the convolution equation $y(n) = h(n) * x(n)$. To find $h(n)$ for the system, we set $x(n)$ equal to $\delta(n)$. In this case, with a characteristic equation of $p + 1 = 0$, and a particular solution of zero, as seen in Table 2.1, we have

$$h(n) = c_1(-1)^n \quad n \geq 0$$

To find c_1, we have from the solution for $h(n)$ that $h(0) = c_1$. Also we can substitute in the given difference equation with $x(n) = \delta(n)$ to find $h(0)$. We will have

$$h(0) + h(-1) = \delta(0)$$

But $h(-1) = 0$ since the output starts at $n = 0$. Thus,

$$h(0) = \delta(0) = 1$$

We then equate $h(0)$ obtained from the solution and $h(0)$ obtained from the given system to get

$$h(0) = c_1 = h(0) = 1$$

$$c_1 = 1$$

And the solution for the impulse response is finally

$$h(n) = (-1)^n \quad n \geq 0$$

Once we have $h(n)$, we can find $y(n)$ using the convolution equation

$$y(n) = x(n) * h(n) = \sum_{m=-\infty}^{+\infty} u(m)(-1)^{n-m} u(n-m) = \sum_{m=0}^{n}(-1)^{n-m} = (-1)^n \sum_{m=0}^{m=n}\left[(-1)^{-1}\right]^m$$

By using the geometric series summation, we simplify to get

$$y(n) = (-1)^n \left[\frac{1-(-1)^{n+1}}{1-(-1)}\right] = \frac{(-1)^n}{2}\left[1-(-1)^{n+1}\right] = \frac{(-1)^n - (-1)^{2n+1}}{2} \quad n \geq 0$$

EOCE 2.9

Consider the following systems:

1. $y(n) + .6y(n-1) = x(n)$
2. $y(n) + y(n-1) + .1y(n-2) = x(n) - x(n-1)$
3. $y(n) + y(n-1) + .1y(n-2) + 8y(n-3) = x(n)$
4. $y(n) - .8y(n-4) = x(n)$

For all systems, use MATLAB to find the impulse responses in the range $-10 \leq n \leq 20$, the step responses and the sinusoidal responses for $x(n) = .5\sin(n)u(n)$.

SOLUTION

As we know, the general difference equation is given as

$$y(n) = \sum_{m=0}^{M} b_m x(n-m) - \sum_{k=1}^{N} a_k y(n-k)$$

In MATLAB, the function called filter is used to solve difference equations for a particular input $x(n)$. The function filter is used by typing

$$y = \text{filter}(b, a, x)$$

where
  y is the output
  b is the input row coefficients associated with the input
  a is the input row coefficients associated with the output y as seen in the previous general equation
  x is the given input

For the first system

$$y(n) + 0.6y(n-1) = u(n)$$

we will enter the b and the a vectors and use the functions stepsignal and impulsesignal defined in Chapter 1 and write the script EOCE 2_9 to find the corresponding plots.

The plots are shown in Figure 2.23. You can see that the system is stable.
    For the second system

$$y(n) + y(n-1) + .1y(n-2) = x(n)$$

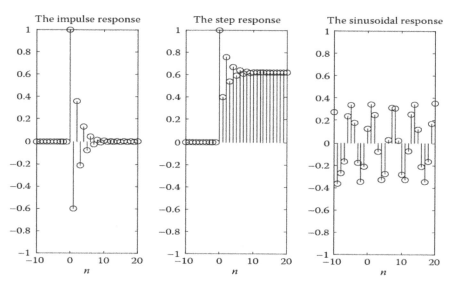

FIGURE 2.23 Signals for EOCE 2.9.

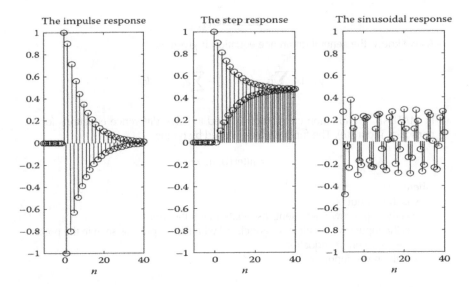

FIGURE 2.24 Signals for EOCE 2.9.

we will use the same MATLAB script given earlier (EOCE 2_9) with the changes.

```
b = [1];
a =[1 1 0.1];
```

The plots are shown in Figure 2.24. The system is stable. We can see that using MATLAB by typing

```
Eigenvalues = roots([1 1 0.1])
```

to get

```
Eigenvalues =
 - 0.8873
 - 0.1127
```

For the third system

$$y(n)+y(n-1)+.1y(n-2)+8y(n-3)=x(n)-x(n-1)$$

we will use the previous script (EOCE 2_9) with the changes

```
b = [1 -1];
a = [1 1 0.1 8];
```

to get the plots shown in Figure 2.25. The system is not stable. We can see that using MATLAB by typing

```
Eigenvalues = roots([1 1 0.1])
```

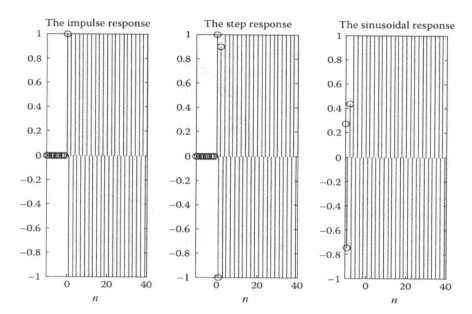

FIGURE 2.25 Signals for EOCE 2.9.

we get

```
Eigenvalues =
- 2.3755
0.6878 + 1.7014i
0.6878 - 1.7014i
```

For the last system

$$y(n) - .8y(n-4) = x(n) + .1x(n-2)$$

we use

```
b = [1 0 0.1];
a = [1 0 0 0 -0.8];
```

The plots are shown in Figure 2.26. The system is stable. That is asserted by typing

```
Eigenvalues = roots([1 0 0 0 -.8])
```

to get

```
Eigenvalues =
- 0.9457
-0.0000 + 0.9457i
-0.0000 - 0.9457i
0.9457
```

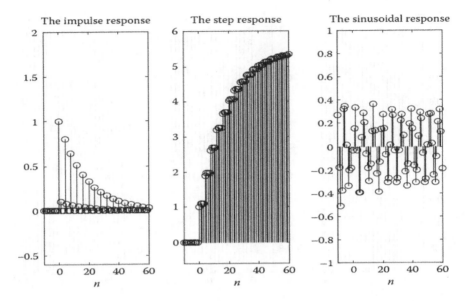

FIGURE 2.26 Signals for EOCE 2.9.

EOCE 2.10

Consider the following system:

$$y(n) + y(n-2) = x(n)$$

1. Find the impulse response.
2. Find the output if $x(n) = u(n)$.

SOLUTION

1. The characteristic equation by inspection is $p^1 + 1 = 0$ with the roots $p_1 = j$ and $p_2 = -j$ or $p_1 = e^{j\pi/2}$ and $p_2 = e^{-j\pi/2}$. Thus,

$$h(n) = c_1 e^{j\pi/2n} + c_2 e^{-j\pi/2n}$$

We must find c_1 and c_2 to completely find $h(n)$. From the solution just obtained, we have

$$h(0) = c_1 + c_2$$

and

$$h(1) = c_1 e^{j\pi/2} + c_2 e^{-j\pi/2} = jc_1 - jc_2$$

We now find the same from the system given to us with the input being the impulse signal and get

$$h(0) + h(-2) = \delta(0) = 1$$

and

$$h(1) + h(-1) = \delta(1) = 0$$

We now equate the $h(0)$ and the $h(1)$ obtained from the solution we arrived at with $h(0)$ and $h(1)$ obtained from the difference equation that was given to us. We will have the two simultaneous algebraic equations

$$c_1 + c_2 = 1$$

$$jc_1 - jc_2 = 0$$

The result will be

$$c_1 = c_2 = \frac{1}{2}$$

and the impulse response is then

$$h(n) = \frac{1}{2}\left[e^{j\pi/2n} + e^{-j\pi/2n} \right] = \cos\left(\frac{\pi}{2} n \right) n \geq 0$$

2. Using convolution, the output $y(n)$ is

$$y(n) = h(n) * x(n) = \sum_{m=-\infty}^{+\infty} u(m) \cos\left(\frac{\pi}{2} n - \frac{\pi}{2} m \right) u(n - m)$$

$$= \sum_{m=0}^{n} \cos\left(n\frac{\pi}{2} - m - \frac{\pi}{2} \right)$$

Using trigonometric identities, we arrive at

$$y(n) = \sum_{m=0}^{n} \cos\left(n\frac{\pi}{2} \right) \cos\left(m\frac{\pi}{2} \right) + \sin\left(n\frac{\pi}{2} \right) \sin\left(m\frac{\pi}{2} \right)$$

$$= \cos\left(n\frac{\pi}{2} \right) \sum_{m=0}^{n} \cos\left(m\frac{\pi}{2} \right) + \sin\left(n\frac{\pi}{2} \right) \sum_{m=0}^{n} \sin\left(m\frac{\pi}{2} \right)$$

EOCE 2.11

So far in this chapter, we have learned that a discrete linear system can be represented by its block diagram, its difference equation, and its impulse response $h(n)$. Given in the following are systems represented as difference equations. Find the other two representations.

1. $y(n) + 2y(n - 1) = x(n)$

2. $y(n) - 4y(n-2) = x(n)$
3. $y(n) - .9y(n-1) + 10y(n-2) + y(n-3) = x(n) + x(n-1)$

SOLUTION

1a. *The block diagram representation*
   Let $y(n-1) = Dy$ and $y(n-2) = D^2y$, and so on. The difference equation can be written then as

$$y(n) + 2Dy(n) = x(n)$$

Solve for $y(n)$ to get

$$y(n) = -2Dy(n) + x(n)$$

The system is first order in y and should have one delay element in the block diagram. As we did earlier, we will have a summer before and after each delay, we will feed $-2y(n)$ to the summer before the delay and $x(n)$ to the summer after the delay. The block diagram is shown in Figure 2.27.

1b. *The impulse response representation*
   The impulse response representation completely describes the discrete system since, given any input we can find the output as $y = x(n) * h(n)$. We start by letting $x(n) = \delta(n)$ in the difference equation given

$$y(n) + 2y(n-1) = \delta(n)$$

The character equation is $p + 2 = 0$ or $p = -2$ is the root. Thus,

$$h(n) = c_1(-2)^n \quad n \geq 0$$

To find c_1, we have

$$h(0) = c_1(-2)^0 c_1$$

and $h(0)$ in the given difference equation is found as

$$h(0) + 2h(-1) = \delta(0) = 1$$

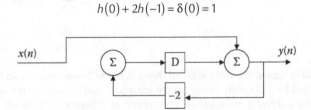

FIGURE 2.27   System for EOCE 2.11.

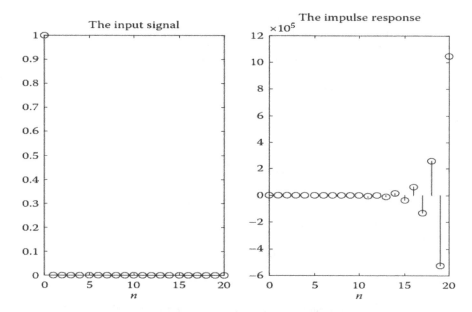

FIGURE 2.28   Signals for EOCE 2.11.

So $h(0) = 1 = c_1$ and finally

$$h(n) = 1(-2)^n \; n \geq 0$$

This impulse response can be found by using MATLAB as in the script EOCE2_111.
The plot is shown in Figure 2.28.

2a. *The block diagram representation*
The difference equations can be written as

$$y(n) = 4D^2 y(n) + x(n)$$

We need two delays in this case and the block diagram is shown in Figure 2.29.

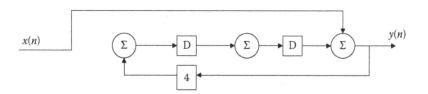

FIGURE 2.29   System for EOCE 2.11.

2b. *The impulse response representation*
The difference equation is

$$y(n) - 4y(n-2) = \delta(n)$$

The character equation is

$$p^2 - 3 = 0 \quad \text{or} \quad p^2 = 3$$

$$p = \pm\sqrt{4} = \pm 2$$

The impulse response is then

$$h(n) = c_1(2)^n + c_2(-2)^n$$

From this last equation for $h(n)$, we get

$$h(0) = c_1 + c_2$$

$$h(1) = 2c_1 - 2c_2$$

and so we need values for $h(0)$ and $h(1)$. We do that using the given difference equation and write

$$h(0) - 4h(-2) = \delta(0) = 1$$

and

$$h(1) - 4h(-1) = \delta(1) = 0$$

Therefore, we have now the two algebraic equations to solve

$$1 = c_1 + c_2$$

$$0 = 2c_1 - 2c_2$$

The results are

$$c_1 = c_2 = \frac{1}{2}$$

and the final solution is then

$$h(n) = \frac{1}{2}\left[(2)^n + (-2)^n\right] n \geq 0$$

$h(n)$ can also be found using MATLAB as in the script EOCE2_112.
The plot is shown in Figure 2.30.

3a. *The block diagram representation*
The difference equation can be written as

$$y(n) = 0.9Dy(n) - 10D^2y(n) - D^3y(n) + x + Dx(n)$$

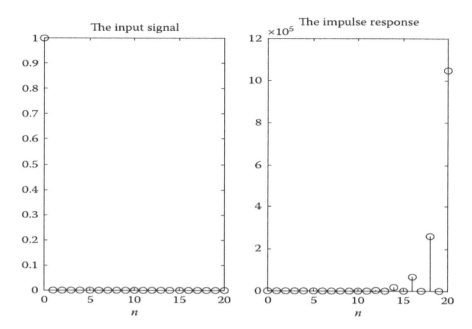

FIGURE 2.30   Signals for EOCE 2.11.

Here we need three delay elements since the system is third order. The block diagram is shown in Figure 2.31.

3b. *The impulse response representation*

We will use MATLAB to find the impulse response as in the script EOCE2_113.

The plot is shown in Figure 2.32. The system is unstable since the magnitude of some of the roots is bigger than 1. The MATLAB command

```
Eigenvalues = roots([1 -.9 10 1])
will result in
EigenValues =
0.4995 + 3.1384i
0.4995 - 3.1384i
- 0.0990
```

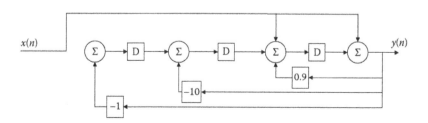

FIGURE 2.31   System for EOCE 2.11.

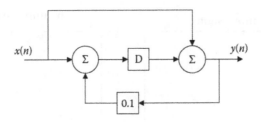

FIGURE 2.32   Signals for EOCE 2.11.

EOCE 2.12

Consider the block diagrams shown in Figures 2.33–2.36. Find the other representations.

SOLUTION

For the system in Figure 2.33, the output after the first summer is $x(n) + 0.1y(n)$. The output at the second summer is $x(n-1) + 0.1y(n-1) + x(n)$. The output of the second summer is also $y(n)$. Therefore,

$$y(n) = 0.1y(n-1) + x(n) + x(n-1)$$

The impulse response is obtained graphically using MATLAB as in the script EOCE2_121.

The plot is shown in Figure 2.37.

For the system in Figure 2.34, the output after the first summer is $x(n)$. After the second summer, the output is $x(n-1) + x(n)$. After the third summer, the output is $x(n-2) + x(n-1) + x(n)$. This output is actually $y(n)$. Therefore,

$$y(n) = x(n) + x(n-1) + x(n-2)$$

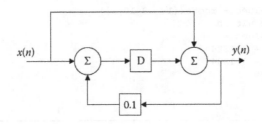

FIGURE 2.33   System for EOCE 2.12.

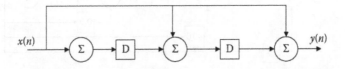

FIGURE 2.34   System for EOCE 2.12.

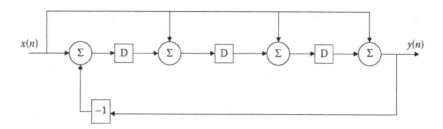

FIGURE 2.35 System for EOCE 2.12.

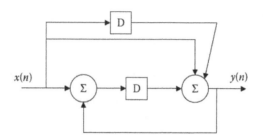

FIGURE 2.36 System for EOCE 2.12.

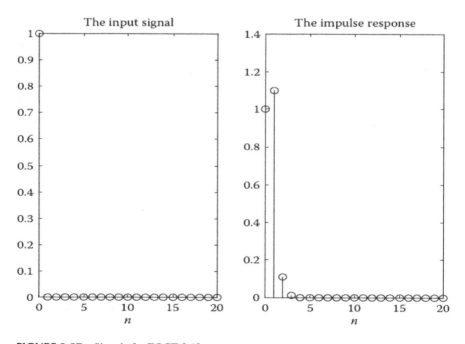

FIGURE 2.37 Signals for EOCE 2.12.

To calculate the impulse responses we simply let $x(n) = \delta(n)$. Therefore,

$$h(n) = \delta(n) + \delta(n-1) + \delta(n-2)$$

We can also use MATLAB to find the impulse response as in the script EOCE2_122.
The plot is shown in Figure 2.38.

For the system in Figure 2.35, the output of the first summer is $x(n) + y(n)$. At the output of the second summer, we have $x(n) + x(n-1) + y(n-1)$. At the output of the third summer, we have $x(n) + x(n-1) + x(n-2) + y(n-2)$. At the output of the last summer, we have

$$y(n) = x(n) + x(n-1) + x(n-2) + x(n-3) + y(n-3)$$

and the difference equation is

$$y(n) - y(n-3) = x(n) + x(n-1) + x(n-2) + x(n-3)$$

The impulse response can be evaluated using MATLAB as in the script EOCE2_123.
The plot is shown in Figure 2.39.

For the system shown in Figure 2.36, the output of the first summer is $x(n) + y(n)$. At the output of the second summer, we have

$$y(n) = x(n) + 2x(n-1) + y(n-1)$$

The impulse response using MATLAB is obtained using the script EOCE2_124.
The plot is shown in Figure 2.40.

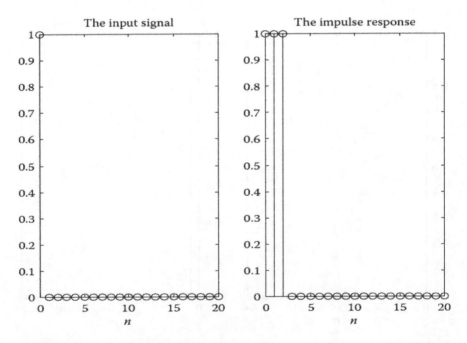

FIGURE 2.38   Signals for EOCE 2.12.

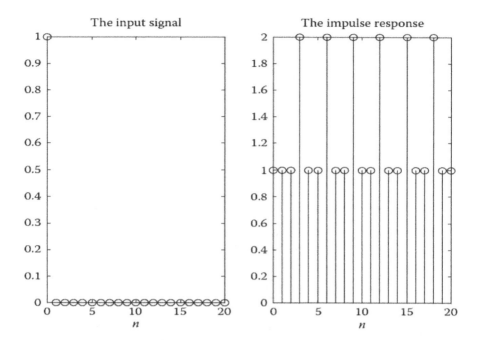

FIGURE 2.39   Signals for EOCE 2.12.

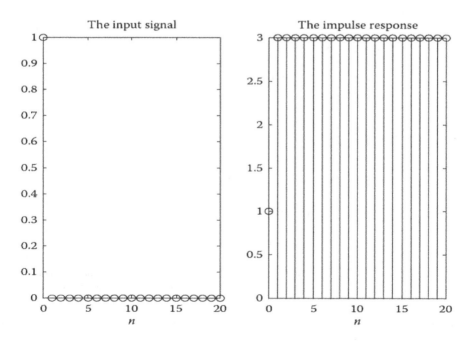

FIGURE 2.40   Signals for EOCE 2.12.

We can also find this response analytically. Recall the general form of the difference equation:

$$y(n) + \sum_{k=1}^{N} a_k y(n-k) = \sum_{k=0}^{M} b_k x(n-k)$$

If N, the degree of the difference equation, is less than M, the general form of $h(n)$ is

$$h(n) = \sum_{k=1}^{N} c_k (p_k)^n + \sum_{k=0}^{M} A_k \delta(n-k)$$

where p_k is the k-th root of the characteristic equation of the system. In this example, $N = 1$ and $M = 1$. Therefore,

$$h(n) = \sum_{k=1}^{1} c_k (P_k)^n + \sum_{k=0}^{0} A_k \delta(n-k) = c_1 (p_1)^n + A_0 \delta(n)$$

We need to find the constant c_1 and A_0. We do that by evaluating the equation for $h(n)$ at $n = 0$ and $n = 1$ to get

$$h(0) = c_1 + A_0$$

$$h(1) = p_1 c_1$$

But p_1 is the characteristic root of the characteristic equation and is 1. Thus,

$$h(0) = c_1 + A_0$$

$$h(1) = c_1$$

To solve for c_1 and A_0, we evaluate the difference equation at $n = 0$ and $n = 1$ when $x(n) = \delta(n)$. We will get

$$h(0) - h(-1) = \delta(0) + 2\delta(-1)$$

which gives $h(0) = 1$. We also have

$$h(1) - h(0) = \delta(1) + 2\delta(0)$$

which gives $h(1) = 3$. By comparing this $h(1)$ with the $h(1)$ obtained earlier, we get $c_1 = 3$ and $A_0 = h(0) - c_1 = 1 - 3 = -2$. Therefore, the final impulse response is

$$h(n) = 3(1)^n - 2\delta(n) = 3 - 2\delta(n)$$

which agrees with the plot in Figure 2.40.

EOCE 2.13

Consider the discrete systems represented by the impulse responses.

1. $h(n) = (.3^n)u(n)$
2. $h(n) = ((.3)^n + (.2)^n)u(n)$
3. $h(n) = \delta(n) + \delta(n-1) + \delta(n-2) + \delta(n-3)$

Find the difference equations first then use MATLAB to find the output $y(n)$ if $x(n) = u(n) - u(n-5)$.

SOLUTION

1. For the first system, multiply $h(n) = (.3)^n u(n)$ by .3 and shift by 1 to get

$$(.3)h(n-1) = (.3)(.3)^{n-1}u(n-1)$$

If we subtract these last two equations, we will have

$$h(n) - .3h(n-1) = (.3)^n u(n) - (.3)(.3)^{n-1}u(n-1)$$

We can rearrange terms and write

$$h(n) - .3h(n-1) = (.3)^n u(n) - (.3)^n u(n-1) = (.3)^n \left[u(n) - u(n-1)\right]$$

But $[u(n) - u(n-1)$ is the impulse at $n = 0$. Thus,

$$h(n) - .3h(n-1) = (.3)^n \delta(n) = \delta(n)$$

since $(.3)^n \delta(n)$ has value only when $n = 0$. We know that if the input $x(n)$ is $\delta(n)$, then the output is $y(n) = h(n)$. Therefore, the difference equation is

$$y(n) - .3y(n-1) = x(n)$$

At this point, we have the difference equation and the input, so we can use the MATLAB filter function to find $y(n)$. We will use the MATLAB functions stepsignal and x1plusx2 to generate $u(n)$ and $-u(n-1)$ and then add them to get $u(n) - u(n-1)$. We do that because these two step signals start at different indices. The MATLAB script EOCE2_131 will be used. The plots are shown in Figure 2.41.
2. For the second system

$$h(n) = (.3)^n u(n) + (.2)^n u(n)$$

This impulse response can be thought of as a sum of the output responses each due to one system as shown in Figure 2.42. We have

$$h_1(n) = (.3)^n u(n)$$

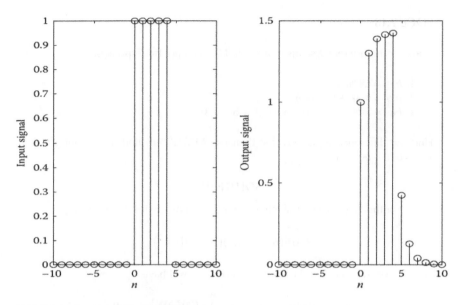

FIGURE 2.41   Signals for EOCE 2.13.

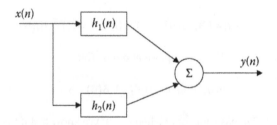

FIGURE 2.42   System for EOCE 2.13.

Multiply by .3 and shift by 1 to get

$$.3h_1(n-1) = (.3)(.3)^{n-1}u(n-1)$$

By subtracting the last two equations we get

$$h_1(n) - .3h_1(n-1) = (.3)^n u(n) - (.3)^n u(n-1) = (.3)^n \left[u(n) - u(n-1) \right]$$

$$= (.3)^n \delta(n) = \delta(n)$$

Therefore, the difference equation is

$$y_1(n) - .3y_1(n-1) = x(n)$$

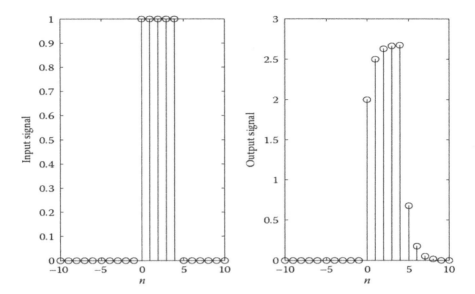

FIGURE 2.43   System for EOCE 2.13.

Similarly, $h_2(n) = (.2)^n u(n)$ can be shown to have the difference equation representation

$$y_2(n) - .2y_2(n-1) = x(n)$$

We will use MATLAB to find $y(n) = y_1(n) + y_2(n)$ for $x(n) = u(n) - u(n-5)$ by writing the EOCE2_132 script.
The plots are shown in Figure 2.43.
3. For the third system,

$$h(n) = \delta(n) + \delta(n-1) + \delta(n-2) + \delta(n-3)$$

We know that when $x(n) = \delta(n)$, then $y(n) = h(n)$. The difference equation is then

$$y(n) = x(n) + x(n-1) + x(n-2) + x(n-3)$$

The output is found using MATLAB as in the EOCE2_133 script.
The plots are shown in Figure 2.44.

EOCE 2.14

Consider the following systems shown in Figures 2.45 and 2.46. Let $x(n) = u(n)$ in Figure 2.45 and $x(n) = \delta(n)$ in Figure 2.46. Find $y(n)$ for both cases. Let

$$h_1(n) = h_2(n) = h_3(n) = (.5)^n u(n)$$

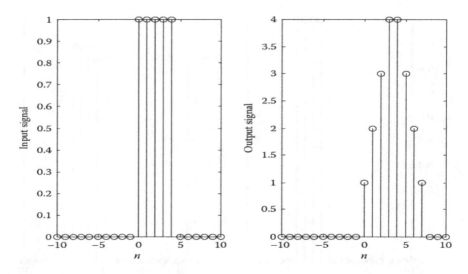

FIGURE 2.44　Signals for EOCE 2.13.

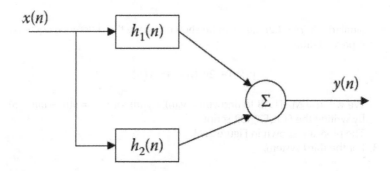

FIGURE 2.45　System for EOCE 2.14.

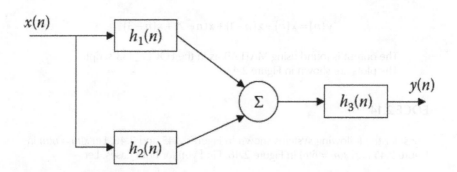

FIGURE 2.46　System for EOCE 2.14.

SOLUTION

For the system in Figure 2.45, the output $y(n)$ is

$$y(n) = x(n) * h_1(n) + x(n) * h_2(n)$$

and

$$x(n) * h_1(n) = \sum_{m=-\infty}^{+\infty} x(m) h_1(n-m) = \sum_{m=-\infty}^{+\infty} u(m)(.5)^{n-m} u(n-m)$$

$$= (.5)^n \sum_{m=0}^{n} \left((.5)^{-1}\right)^m = (.5)^n \frac{\left[1 - \left((.5)^{-1}\right)^{n+1}\right]}{1 - (.5)^{-1}}$$

$$= -(.5)^n + (.5)^{-1} = 2 - (.5)^n \quad n \geq 0$$

Similarly

$$x(n) * h_2(n) = 2 - (.5)^n \quad n \geq 0$$

Therefore, the final output is

$$y(n) = 4 - 2(.5)^n \quad n \geq 0$$

For the system in Figure 2.46, the output of the summer is

$$x(n) * h_1(n) + x(n) * h_2(n) = z(n)$$

The output of the system is

$$y(n) = z(n) * h_3(n)$$

We have

$$x(n) * h_1(n) = 2 - (.5)^n \quad n \geq 0$$

and

$$x(n) * h_2(n) = 2 - (.5)^n \quad n \geq 0$$

The output of the summer is

$$z(n) = 4 - 2(.5)^n \quad n \geq 0$$

and the output of the whole system is then

$$y(n) = \sum_{m=-\infty}^{+\infty} z(m)h_3(n-m) = \sum_{m=-\infty}^{+\infty} \left(4 - 2(.5)^m\right)u(m)(.5)^{n-m}u(n-m)$$

$$y(n) = 4(.5)^n \sum_{m=0}^{n} (.5^{-1})^m - 2(.5)^n \sum_{m=0}^{n} 1$$

Finally, the output is

$$y(n) = 4(.5)^n \frac{\left[1 - \left((.5)^{-1}\right)^{n+1}\right]}{1 - (.5)^{-1}} - 2(.5)^n [n+1] = 8 - 4(.5)^n - 2(.5)^n [n+1] \ n \geq 0$$

EOCE 2.15

Consider the following signals:

$$x_1(n) = u(n) \quad 0 \leq n \leq 9$$

$$x_2(n) = u(n) \quad 0 \leq n \leq 18$$

$$x_3(n) = u(n) \quad -4 \leq n \leq 5$$

$$x_4(n) = u(n) \quad -9 \leq n \leq 0$$

Correlate $x_1(n)$ with itself and all other signals.

SOLUTION

We will use the MATLAB script EOCE2_15 to do this. We can see clearly that for $R_{x_1x_1}$ the strongest correlation is when the lag k is equal to zero. In this case $R_{x_1x_1}$ at $k = 0$ is obtained by summing 1(1) 10 times to get 10 as the strength. For $R_{x_1x_2}$ the maximum strength is at $k = -9$. This is true because for a full overlap we need to shift $x_2(n)$ in the reverse direction, which means negative k. If $k = 0$ in this case we have only one sample overlap: 1(1) = 1. This is evident from the plot shown in Figure 2.47.

EOCE 2.16

One important application of cross-correlation is the estimation of the impulse response $h(n)$. Given a system with impulse response $h(n)$, the output using

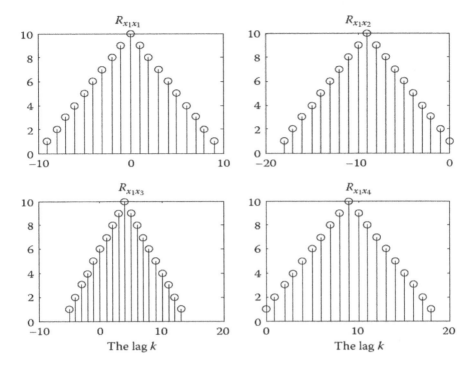

FIGURE 2.47 Plots for EOCE 2.15.

convolution is $y(n) = x(n) * h(n)$. If we cross-correlate the output of the system with a noisy input with normal distribution, then the cross-correlation

$$R_{yr}(k) = y(n) * r(-n) = [r(n) * h(n)] * r(-n) = R_{rr}(k) * h(n)$$

where $r(n)$ is a noisy random input; in the previous equation we have used the fact that $r(n) * h(n) = h(n) * r(n)$. The signal $R_{rr}(k)$ is the auto-correlation of the input noise. If the input noise has a bandwidth that is much greater than the bandwidth of the system, we write

$$R_{yr}(k) \approx h(n)$$

The reason is that if the auto-correlation of the input noise is approximately an impulse, then we know that it contains a very wide band of frequencies. To see that let us look at the noise auto-correlation. We will use MATLAB script EOCE2_161 to do that.

The plot is shown in Figure 2.48. You can easily see that the auto-correlation is approximately the same as the impulse signal, which has a huge bandwidth. Therefore, we say that if we cross-correlate the random normally distributed input noise with the output of the system, we will have an approximation of the impulse response of the system.

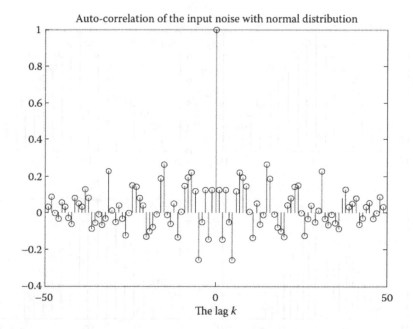

FIGURE 2.48 Plot for EOCE 2.16.

Let us look at an example. Consider the following system:

$$y(n) + .5y(n-1) = x(n)$$

SOLUTION

We will plot its actual impulse response and then the approximation to it using cross-correlation between the output and the input noise. We do that using MATLAB script EOCE2_162.

The plots are shown in Figure 2.49.

2.21   END-OF-CHAPTER PROBLEMS

EOCP 2.1

Consider the following systems for $n \geq 0$:

1. $y(n) = nx(n)$
2. $y(n) = \cos(x(n)) + u(n)$
3. $y(n) = \sqrt{nx}(n)$
4. $y(n) = (4)^n \sin(n)$
5. $y(n) = \cos\left(n\dfrac{\pi}{2} - 1\right)$

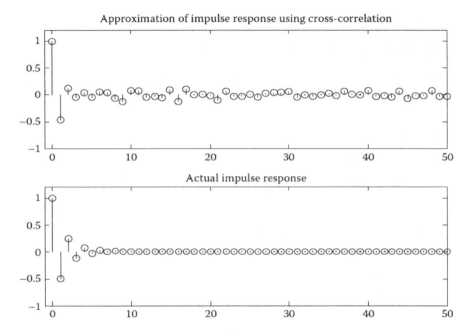

FIGURE 2.49   Plots for EOCE 2.16.

6. $y(n) = \dfrac{\sin(nx(n))}{n}$

7. $y(n) = \sin(n)\cos(x(n))$

8. $y(n) = y(n-1) + x(n)$

9. $y(n) = ny(n-1) + x(n)$

10. $y(n) = y(n-1) + y(n-2) + x(n)$

Are the aforementioned systems linear? Are they time-invariant systems? Show work.

EOCP 2.2

Consider the following systems:

1. $h(n) = u(n)$, $x(n) = u(n)$
2. $h(n) = (.2)^n u(n)$, $x(n) = \delta(n)$
3. $h(n) = (.3)^n u(n)$, $x(n) = u(n)$
4. $h(n) = (.3)^n u(n)$, $x(n) = (.2)^n u(n)$
5. $h(n) = (.3)^n nu(n)$, $x(n) = \delta(n)$
6. $h(n) = n(.5)^n \cos\left(\dfrac{\pi n}{2} + 1\right)u(n)$, $x(n) = \delta(n)$
7. $h(n) = (.4)^n u(n)$, $x(n) = u(n) - u(n-2)$

8. $h(n) = u(n) - u(n-2)$, $x(n) = u(n) - u(n-2)$
9. $h(n) = (.4)^n\, u(n)$, $x(n) = (.5)^n\,[u(n) - u(n-1)]$
10. $h(n) = (.1)^n\, u(n)$, $x(n) = (.5)^n\,[u(n) - u(n-5)]$

Find the output $y(n)$ for each system using the convolution sum equation. Are the systems stable?

EOCP 2.3

Consider the following finite signals:

1. $x_1(n) = \left\{ \begin{matrix} 1 & 1 & 1 & 1 \\ \uparrow & & & \end{matrix} \right\}, x_2(n) = \left\{ \begin{matrix} 1 & 1 & 1 & 1 \\ \uparrow & & & \end{matrix} \right\}$

2. $x_1(n) = \left\{ \begin{matrix} 1 & 1 & 1 & 1 \\ & & \uparrow & \end{matrix} \right\}, x_2(n) = \left\{ \begin{matrix} 1 & 1 & 1 & 1 \\ & & \uparrow & \end{matrix} \right\}$

3. $x_1(n) = \left\{ \begin{matrix} -1 & 2 & 1 & 3 \\ & & \uparrow & \end{matrix} \right\}, x_2(n) = \left\{ \begin{matrix} 1 & 2 \\ \uparrow & \end{matrix} \right\}$

4. $x_1(n) = \left\{ \begin{matrix} -1 & -2 & -3 & -4 \\ & & \uparrow & \end{matrix} \right\}, x_2(n) = \left\{ \begin{matrix} 1 & 2 \\ \uparrow & \end{matrix} \right\}$

5. $x_1(n) = \left\{ \begin{matrix} 1 & -1 & 2 & -2 & 3 & -3 \\ & & & \uparrow & & \end{matrix} \right\}, x_2(n) = \left\{ \begin{matrix} 1 & 2 & 3 \\ & \uparrow & \end{matrix} \right\}$

Find $x(n) = x_1(n) * x_2(n)$ for each case given earlier.

EOCP 2.4

Consider the following systems with the initial conditions:

1. $y(n) - .6y(n-1) = 0$, $y(-1) = 1$
2. $y(n) - .6y(n-2) = 0$, $y(-1) = 0$, $y(-2) = 1$
3. $y(n) - .6y(n-1) + .6y(n-2) = 0$, $y(-1) = 0$, $y(-2) = -1$
4. $y(n) - .1y(n-3) = 0$, $y(-1) = y(-2) = 0$, $y(-3) = 1$
5. $y(n) + .1y(n-1) + y(n-3) = 0$, $y(-1) = 0$, $y(-2) = 2$, $y(-3) = 0$
6. $y(n) + .6y(n-2) = 0$, $y(-1) = y(-2) = 1$

Are the aforementioned systems stable? What is the output due to the given initial conditions?

EOCP 2.5

Consider the following systems:

1. $y(n) - .6y(n-1) = \sin(n)$, $y(-1) = 0$
2. $y(n) - .6y(n-2) = \delta(n)$, $y(-1) = y(-2) = 0$
3. $y(n) - .6y(n-1) + .6y(n-2) = 3\delta(n)$, $y(-1) = 0$, $y(-2) = 0$
4. $y(n) - .1y(n-3) = \delta(n)$, $y(-1) = y(-2) = y(-3) = 0$

5. $y(n) + .1y(n - 1) + y(n - 3) = 2\delta(n)$, $y(-1) = y(-2) = y(-3) = 0$
6. $y(n) + .6y(n - 2) = 3u(n)$, $y(-1) = y(-2) = 0$

Find the total output, $y(n)$, for each system mentioned earlier.

EOCP 2.6

Consider the following difference equations:

1. $y(n) - .6y(n - 1) = x(n)$
2. $y(n) - .6y(n - 2) = x(n) - x(n - 1)$
3. $y(n) - .6y(n - 1) + .6y(n - 2) = x(n)$
4. $y(n) - .1y(n - 3) = x(n) + x(n - 1)$
5. $y(n) + .1y(n - 1) + y(n - 3) = x(n)$
6. $y(n) + .6y(n - 2) = x(n - 2)$

Draw the block diagram for each system.

EOCP 2.7

Consider the following systems:

1. $h(n) = \delta(n) - \delta(n - 1)$
2. $h(n) = \delta(n) - \delta(n - 1) + \delta(n - 2) + \delta(n - 4)$
3. $h(n) = u(n)$
4. $h(n) = (.3)^n u(n)$
5. $h(n) = (.3)^n u(n) + (.6)^n u(n)$
6. $h(n) = u(n)$, $0 \le n \le 5$

What are the difference equations that represent these systems?

EOCP 2.8

Consider the following difference equations:

1. $y(n) + 3y(n - 1) = x(n)$
2. $y(n) + 3y(n - 1) + 3y(n - 2) = x(n)$
3. $y(n) - .1y(n - 2) = x(n)$
4. $y(n) + .1y(n - 2) = x(n)$
5. $y(n) + 3y(n - 1) = x(n) + x(n - 1)$
6. $y(n) + 3y(n - 1) + 3y(n - 2) = x(n) + x(n - 1) + x(n - 2)$
7. $y(n) - .1y(n - 2) = x(n) + x(n - 1) + x(n - 2)$
8. $y(n) + .1y(n - 2) = x(n) + x(n - 1) + x(n - 2) + x(n - 4)$
9. $y(n) + 3y(n - 1) = x(n) + x(n - 1) + x(n - 2)$
10. $y(n) + 3y(n - 1) = x(n) + x(n - 1) + x(n - 2) + x(n - 3)$

What are the impulse responses for each system mentioned earlier?

EOCP 2.9

Consider the block diagrams shown in Figures 2.50–2.59. What are the impulse responses for each system?

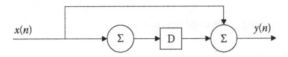

FIGURE 2.50   System for EOCP 2.9.

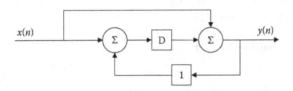

FIGURE 2.51   System for EOCP 2.9.

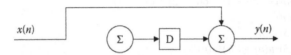

FIGURE 2.52   System for EOCP 2.9.

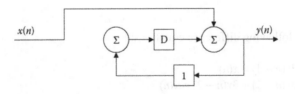

FIGURE 2.53   System for EOCP 2.9.

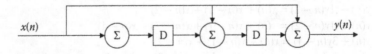

FIGURE 2.54   System for EOCP 2.9.

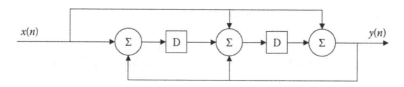

FIGURE 2.55 System for EOCP 2.9.

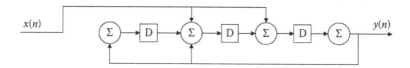

FIGURE 2.56 System for EOCP 2.9.

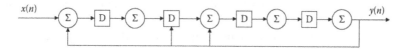

FIGURE 2.57 System for EOCP 2.9.

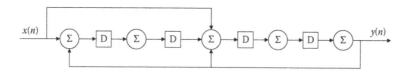

FIGURE 2.58 System for EOCP 2.9.

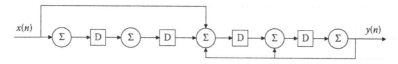

FIGURE 2.59 System for EOCP 2.9.

EOCP 2.10

Consider the following systems:

1. $y(n) + ky(n-1) = x(n)$
2. $y(n) + ky(n-1) + y(n-2) = x(n)$
3. $y(n) - ky(n-1) = x(n)$

4. $y(n) + y(n-1) + ky(n-2) = x(n)$
5. $y(n) + .1y(n-1) + .3y(n-2) + y(n-3) = x(n)$
   a. For what values of k are the systems stable?
   b. Use MATLAB to find the step, impulse, and the sinusoidal responses for all systems with suitable k.

EOCP 2.11

Consider the following difference equation:

$$y(n) - yk(n-1) + 0.5y(n-2) = 0.5x(n)$$

1. Is the system stable? For what k values?
2. Draw the block diagram representation of the system.
3. Use MATLAB to find the step and the impulse responses for a certain k.
4. Find the impulse response analytically with k of your choice.
5. Find the step response analytically assuming zero initial conditions with suitable k.
6. Solve the difference equation by iterations for $n = 0, 1, 2, 3,$ and 5, if $x(n) = u(n)$ for suitable k.

EOCP 2.12

Consider the following system represented by the difference equation:

$$y(n) - 6y(n-1) + 9ky(n-2) = 19x(n)$$

1. Use MATLAB to find the step and the impulse responses for a k value that makes the system stable.
2. Use MATLAB to find the output $y(n)$ if $x(n) = 2u(n) + 5nu(n)$.
3. Draw the block diagram for the system.

EOCP 2.13

Consider the following systems:

1. $y(n) + .2y(n-1) = x(n)$
2. $y(n) + .2y(n-1) + .1y(n-2) = x(n)$

Approximate the impulse response using cross-correlation. Compare with the actual impulse response.

3 Fourier Series and the Fourier Transform of Discrete Signals

3.1 INTRODUCTION

According to Joseph Fourier, a discrete periodic signal can be represented as a sum of complex exponentials or sinusoids. Also, a nonperiodic discrete signal and a finite-duration discrete signal can be represented as a finite sum of complex exponentials. This last representation for finite-duration discrete signals is called the Fourier transform of discrete signals. There are many advantages for this representation. The most important one is that if the discrete signal is put in the frequency domain, every single frequency in the signal will be clearly identified. Consider the plot of the discrete signal

$$y(n) = \cos(2\pi n) + 200 \cos(4\pi n) + 0.1 \cos(10\pi n)$$

as shown in Figure 3.1. By looking at the plot, you would hardly think that this signal is composed of many sinusoids. You will probably reason that this signal contains no sinusoids at all. In the frequency domain, if we look at the magnitude of the signal, we see something similar to Figure 3.2.

You can clearly see the three frequencies in the figure. You may also argue that these signals of small magnitudes, like the term $0.1 \cos(10\pi n)$, can be neglected. You may be wrong, especially if this term is the term that we are interested in. Before we present discussion about the discrete Fourier series and the discrete Fourier transform, we provide a very useful review on complex numbers.

3.2 REVIEW OF COMPLEX NUMBERS

This is a good place to give a brief review of complex numbers because the chapters ahead, as well as this chapter, are heavily involved with their arithmetic manipulation.

3.2.1 DEFINITION

By definition, the complex number j is the square root of −1. In general, the complex number $C = P + jQ$ consists of two parts: the real part P and the imaginary part Q. C can be represented in many forms as we will see later. If we are multiplying or dividing complex numbers, we would prefer to use the polar form. If we are adding or subtracting complex numbers, we would rather use the rectangular form. The reason for that is the ease each form provides in the corresponding calculation.

Consider two complex numbers C_1 and C_2, where $C_1 = P_1 + jQ_1$ and $C_2 = P_2 + jQ_2$. These complex numbers are given in the rectangular form.

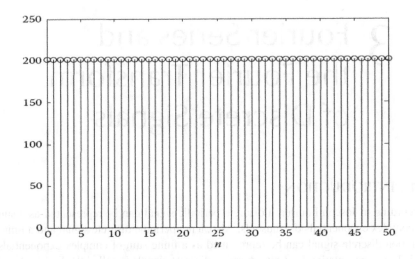

FIGURE 3.1 A discrete sinusoidal signal.

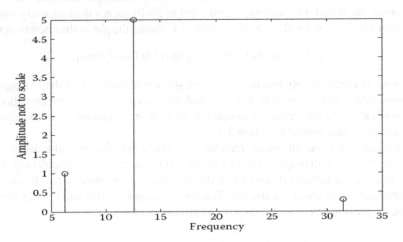

FIGURE 3.2 A discrete signal in the frequency domain.

3.2.2 ADDITION

When we add two complex numbers, we add their real parts and their imaginary parts together to form the addition as follows:

$$C_1 + C_2 = (P_1 + P_2) + j(Q_1 + Q_2) \tag{3.1}$$

3.2.3 SUBTRACTION

When we subtract two complex numbers, we subtract their real parts and their imaginary parts to form the subtraction as follows:

$$C_1 - C_2 = (P_1 - P_2) + j(Q_1 - Q_2) \tag{3.2}$$

3.2.4 MULTIPLICATION

Let us now consider the two complex numbers in polar form. In polar form, C_1 and C_2 are represented as

$$C_1 = M_1 e^{j\theta_1} \text{ and } C_2 = M_2 e^{j\theta_2}$$

where

$$M_1 = \sqrt{P_1^2 + Q_1^2}$$

$$M_2 = \sqrt{P_2^2 + Q_2^2}$$

and

$$\theta_1 = \tan^{-1}\left[\frac{Q_1}{P_1}\right]$$

$$\theta_2 = \tan^{-1}\left[\frac{Q_2}{P_2}\right]$$

We can also represent the complex numbers C_1 and C_2 as $M_1 < \theta_1$ and $M_2 < \theta_2$, where M_1, M_2, θ_1, and θ_2 are as given earlier. The complex number $M_1 < \theta_1$ is read as a complex number with magnitude M_1 and phase angle θ_1. Complex numbers are easily multiplied in polar form as

$$M_1 e^{j\theta_1} M_2 e^{j\theta_2} = M_1 M_2 e^{j(\theta_1 + \theta_2)} = M_1 M_2 \angle(\theta_1 + \theta_2) \qquad (3.3)$$

If we have more than two complex numbers in polar form to be multiplied, we use the same procedure. We multiply their magnitudes and add their phase angles to form the new product.

3.2.5 DIVISION

To divide two complex numbers, we divide their magnitudes and subtract their phase angles.

$$\frac{C_1}{C_2} = \frac{\left[M_1 e^{j\theta_1}\right]}{\left[M_2 e^{j\theta_2}\right]}$$

$$= \left[\frac{M_1}{M_2}\right] e^{j(\theta_1 + \theta_1)} = \left[\frac{M_1}{M_2}\right] \angle(\theta_1 - \theta_1) \qquad (3.4)$$

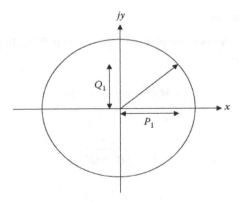

FIGURE 3.3   Rectangular form.

3.2.6   FROM RECTANGULAR TO POLAR

Consider the complex number $C_1 = P_1 + jQ_1$. This representation in the rectangular form is shown in Figure 3.3. To convert to polar form, we write C_1 as $C_1 = M_1 e^{j\theta_1} M_1 < \theta_1$, where

$$M_1 = \sqrt{P_1^2 + Q_1^2} \text{ and } \theta_1 = \tan^{-1}\left[\frac{Q_1}{P_1}\right]$$

3.2.7   FROM POLAR TO RECTANGULAR

Consider the complex number $C_1 = M_1 e^{j\theta_1} = M_1 < \theta_1$ in polar form, as seen in Figure 3.4. To convert to its equivalent rectangular form, we write C_1 as $C_1 = P_1 + jQ_1$, where

$$M_1^2 = P_1^2 + Q_1^2 \text{ and } \tan(\theta_1) = \left[\frac{Q_1}{P_1}\right]$$

These two equations can be solved simultaneously for P_1 and Q_1.

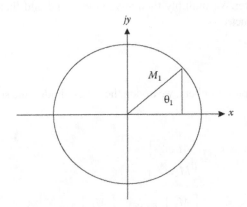

FIGURE 3.4   Polar form.

3.3   FOURIER SERIES OF DISCRETE PERIODIC SIGNALS

We are given the discrete signal $x(n)$ that might be the result of sampling a continuous signal $x(t)$ at $t = nT_s$. This signal is periodic of period N samples if

$$x(n) = x(n + N)$$

for all integer values n. The fundamental period is the smallest N that satisfies the previous condition. The period in seconds would be NT_s, where T_s is the sampling period. In this case, the digital frequency is $\theta = 2\pi/NT_s$. Now consider the complex exponential set

$$\phi_m(n) = e^{jm\theta nT_s} = e^{j2\pi mn/N} \quad m = 0, \pm 1, \pm 2 \cdots \tag{3.5}$$

where

  m is referred to as the frequency index.
  n is the discrete-time index.

Notice that this complex exponential set is periodic with period N for all integers n since

$$\phi_m(n + N) = e^{jm\theta(n+N)T_s} = e^{j2\pi m} e^{j2\pi mn/N} = 1 e^{j2\pi mn/N} = \phi_m(n) \tag{3.6}$$

Notice also that since $\phi_m(n + N) = \phi_m(n)$, there is only one set of complex functions of length N that is unique in the overall set. Unique set means that the members in the set are independent; no one function or complex exponential in the set can be written as a linear combination of the others. In the set $\phi_m(n)$ of length N, we see that

$$\sum_{n=0}^{N-1} \phi_m(n) = \sum_{n=0}^{n-1} e^{j2\pi mn/N} = \begin{cases} N & m = 0, \pm N, \pm 2N, \cdots \\ 0 & m \neq 0, \pm N, \pm 2N, \cdots \end{cases} \tag{3.7}$$

since using the geometric series sum, we have

$$\sum_{n=0}^{N-1} e^{j2\pi mn/N} = \sum_{n=0}^{N-1} \left(e^{j2\pi m/N}\right)^n = \frac{1 - e^{(j2\pi m/N)N}}{1 - e^{j2\pi m/N}} = \begin{cases} N & m = 0, \pm N, \pm 2N, \cdots \\ 0 & m \neq 0, \pm N, \pm 2N, \cdots \end{cases} \tag{3.8}$$

We can also see that

$$\sum_{n=0}^{N-1} e^{j2\pi mn/N} e^{-j2\pi kn/N} = \sum_{n=0}^{N-1} \left(e^{j2\pi n(m-k)/N}\right) = \begin{cases} N & m = k, \pm k + N, \pm k + 2N, \cdots \\ 0 & m \neq k, \pm k + N, \pm k + 2N, \cdots \end{cases} \tag{3.9}$$

This is the orthogonality condition. The periodic signal $x(n)$ can be approximated as the linear combination of the unique set $\phi(m)$ with the N complex exponentials as

$$x(n) = \sum_{n=0}^{N-1} C_m e^{j2\pi mn/N} \tag{3.10}$$

To find the Fourier series coefficients, we multiply the previous approximation by $e^{-j2\pi kn/N}$ and then sum over the period N to get

$$C_m = \frac{1}{N} \sum_{n=0}^{N-1} x(n) e^{-j2\pi mn/N} \tag{3.11}$$

using the geometric series sum and the orthogonality condition we established earlier. Try to see it for yourself.

Writing the approximation to the periodic signal $x(n)$ allows us to clearly see the frequency contents of the signal through the coefficients c_m. These coefficients are complex numbers, yet they have a magnitude and a phase. It can be easily seen that the c_m set is periodic. We write

$$C_{m+N} = \frac{1}{N} \sum_{n=0}^{N-1} x(n) e^{-j2\pi(m+N)n/N} = \frac{1}{N} \sum_{n=0}^{N-1} x(n) e^{-j2\pi n} e^{j2\pi nm/N} = c_m \tag{3.12}$$

Also, if $x(n)$ is real, then the Fourier series magnitude coefficients have even symmetry and the phase of the coefficients have odd symmetry. We finish this topic with an example. Consider the periodic discrete signal $x(n) = \{1\ 0\ 1\ 1\ 0\ 1\ 1\ 0\ 1\ \ldots\}$ with $T_s = 0.1$ s as the sampling interval used to obtain $x(n)$. The period N is clearly 3. Thus, the Fourier series coefficients are calculated as in the following:

$$c_0 = \frac{1}{3} \sum_{n=0}^{2} x(n) e^{-j2\pi 0 n/3} = \frac{1}{3}[x(0) + x(1) + x(2)] = \frac{2}{3}$$

$$c_1 = \frac{1}{3} \sum_{n=0}^{2} x(n) e^{-j2\pi n/3} = \frac{1}{3}\left[x(0) + x(1) e^{-j2\pi/3} + x(2) e^{-j4\pi/3}\right] = \frac{1}{3}[0.5 + j.886]$$

$$c_2 = \frac{1}{3} \sum_{n=0}^{2} x(n) e^{-j4\pi n/3} = \frac{1}{3}\left[x(0) + x(1) e^{-j4\pi/3} + x(2) e^{-j8\pi/3}\right] = \frac{1}{3}[0.5 - j.886]$$

These are the frequency components in the signal $x(n)$. These components are located at $m\theta = m2\pi/NT_s = 20m\pi/3$ rad/s for $m = 0$, 1, and 2. Thus, these frequencies are located at 0, $20\pi/3$, and $40\pi/3$ rads per sample. This is something very difficult to see given only the discrete periodic signal $x(n) = \{1\ 0\ 1\ 1\ 0\ 1\ 1\ 0\ 1\ \ldots\}$

with $T_s = 0.1$ s. We will come back to the Fourier series coefficients in Chapter 5 and see how we can utilize this development to estimate the average power, the average value, and the energy in signals. We will also learn how to compute these Fourier series coefficients using the computer and the very well-known algorithm, the fast Fourier transform.

3.4 DISCRETE SYSTEM WITH PERIODIC INPUTS: THE STEADY-STATE RESPONSE

To start the process, let us consider the input $x(n)$ to be of the following form:

$$x(n) = e^{j\theta n}$$

Notice that the magnitude of $x(n)$ or its phase do not contribute to any information about the input signal frequency and that is why neither were mentioned nor contained in $x(n)$ for simplicity. With $h(n)$, the impulse response, the output of the system in this case is

$$y(n) = x(n) * h(n) = h(n) * x(n) = \sum_{m=-\infty}^{+\infty} h(m)x(n-m) = \sum_{m=-\infty}^{+\infty} h(m)e^{j\theta(n-m)}$$

$$y(n) = \sum_{m=-\infty}^{+\infty} h(m)e^{j\theta n}e^{-j\theta m} = e^{j\theta n}\sum_{m=-\infty}^{+\infty} h(m)e^{-j\theta m}$$

Given that n was taken in the range $-\infty \leq n \leq \infty$, the output here is the steady state output that remains after all the transients die. Finally, with $x(n) = e^{j\theta n}$, we have

$$y_{ss}(n) = x(n)\sum_{m=-\infty}^{+\infty} h(m)e^{-j\theta m}$$

where $y_{ss}(n)$ is the steady-state response and $x(n) = e^{j\theta n}$. Let us define $H(e^{j\theta})$ as

$$H(e^{j\theta}) = \sum_{m=-\infty}^{+\infty} h(m)e^{-j\theta m} \tag{3.13}$$

then

$$y_{ss}(n) = x(n)H(e^{j\theta}) \tag{3.14}$$

Example 3.1

Consider the impulse response for a discrete system as $h(n) = \delta(n) + \delta(n-1)$ and let the input to the system be $x(n) = e^{j2\pi n}$ for $-\infty \le n \le \infty$. Find the steady- state output.

SOLUTION

From the relation $y_{ss}(n) = x(n)H(e^{j\theta})$ with $x(n) = e^{j2\pi n}$ and $H(e^{j\theta})$ as

$$H\left(e^{j\theta}\right) = \sum_{m=-\infty}^{+\infty} \left(\delta(m) + \delta(m-1)\right)e^{-j\theta m} = \left(\delta(0) + \delta(-1)\right)e^{-j\theta(0)} + \left(\delta(1) + \delta(0)\right)e^{-j\theta(1)}$$

$$H\left(e^{j\theta}\right) = 1 + e^{-j\theta}$$

Thus, the steady-state output becomes

$$y_{ss}(n) = e^{j2\pi n}\left(1 + e^{-j\theta}\right) \quad -\infty \le n \le \infty$$

The input is applied at $\theta = 2\pi$. So

$$H\left(e^{j\theta}\right) = 1 + e^{-j2\pi} = 1 + \cos 2\pi - j \sin 2\pi = 2$$

and

$$y_{ss}(n) = 2e^{j2\pi n} \quad -\infty \le n \le +\infty$$

Example 3.2

For the same system as in Example 3.1, and with the input $x(n) = 10 + \cos(n\pi)$, what is the steady-state output?

SOLUTION

To use $y_{ss}(n) = e^{j\theta n}H(e^{j\theta})$, we must put all the terms in $x(n)$ in exponential form and then we can use the superposition principle to find

$$y_{ss1}(n) \text{ due to } x_1(n) = 10 = 10e^{jn(0)}$$

$$y_{ss2}(n) \text{ due to } x_2(n) = \frac{1}{2}e^{j\pi n}$$

$$y_{ss3}(n) \text{ due to } x_3(n) = \frac{1}{2}e^{-j\pi n}$$

where

$$\cos(n\pi) = \frac{e^{jn\pi} + e^{-j\pi n}}{2}$$

From Example 3.1, we have $H(e^{j\theta}) = 1 + e^{-j\theta}$ and for $x_1(n) = 10e^{j0(0)}$, we have

$$H\left(e^{j\theta}\right) = 2$$

$$y_{ss1}\left(n\right) = 10e^{j\theta n}\left(2\right) = 20$$

For $x_2(n) = e^{j\pi n}/2$ with $0 = \pi$, we have

$$H\left(e^{j\theta}\right) = 1 + e^{-j\pi} = 1 + \cos\pi - j\sin\pi = 0$$

and

$$y_{ss2}\left(n\right) = \frac{e^{j\pi n}}{2}\left(0\right) = 0$$

For

$$x_3\left(n\right) = \frac{e^{-j\pi n}}{2} \text{ with } \theta = -\pi$$

we have

$$H\left(e^{j\theta}\right) = 1 + e^{-j\theta} = 1 + e^{j\pi} = 1 + \cos\pi + j\sin\pi = 0$$

and

$$y_{ss3}\left(n\right) = \frac{e^{-j\pi n}}{2}\left(0\right) = 0$$

Therefore, the total solution is the combination as follows:

$$y_{ss}\left(n\right) = y_{ss1}\left(n\right) + y_{ss2}\left(n\right) + y_{ss3}\left(n\right) = 20 + 0 + 0 = 20 \quad -\infty \le n \le +\infty$$

Example 3.3

Find the steady-state output for the system

$$h(n) = (.5)^n u(n)$$

with the input $x(n) = 10$.

SOLUTION

We will find $H(e^{j\theta})$ first.

$$H\left(e^{j\theta}\right) = \sum_{m=-\infty}^{+\infty} h(m)e^{-j\theta m} = \sum_{m=0}^{\infty}(.5)^m e^{-j\theta m} = \sum_{m=0}^{\infty}\left((.5)e^{-j\theta}\right)^m = \frac{1}{1-(.5)e^{-j\theta}}$$

The input is $x(n) = 10 = 10e^{jn0}$. So $\theta = 0$ and

$$H(e^{j\theta}) = \frac{1}{1-(.5)e^{-j0}} = \frac{1}{1-0.5} = 2$$

and the steady-state response is

$$y_{ss}(n) = 10(2) = 20 \quad -\infty \le n \le +\infty$$

3.4.1 GENERAL FORM FOR $y_{ss}(n)$

Consider the input $x(n)$ as

$$x(n) = X\cos(\theta n + \varphi) = \frac{x}{2}\left[e^{j(\theta n+\varphi)} + e^{-j(\theta n+\varphi)}\right]$$

when

    θ is the radian frequency.
    φ is the phase shift.

If $x(n)$ is applied to a linear time-invariant system, the steady-state output using Equation (3.14) will be

$$y_{ss}(n) = \frac{x}{2}e^{j\varphi}e^{j\theta n}H(e^{j\theta}) + \frac{x}{2}e^{-j\varphi}e^{-j\theta n}H(e^{-j\theta})$$

Let us write $H(e^{j\theta})$ in terms of its magnitude and phase as $Me^{j\alpha}$, where M is the magnitude of $H(e^{j\theta})$ and α is its phase shift. The steady-state solution is then

$$y_{ss}(n) = \frac{x}{2}e^{j\varphi}e^{j\theta n}Me^{j\alpha} + \frac{x}{2}e^{-j\varphi}e^{-j\theta n}Me^{-j\alpha}$$

$$= \frac{xm}{2}\left[e^{j\varphi}e^{j(\theta n+\alpha)} + e^{-j\varphi}e^{-j(\theta n+\alpha)}\right]$$

By putting exponents together, we get

$$y_{ss}(n) = \frac{xm}{2}\left[e^{j(\theta n+\alpha+\varphi)} + e^{-j(\theta n+\alpha+\varphi)}\right]$$

and finally the steady-state response is written as

$$y_{ss}(n) = XM \cos(\theta n + \varphi + \alpha) \tag{3.15}$$

Example 3.4

Let $h(n) = \delta(n) + \delta(n-1)$ and let $x(n) = 10 \cos(n(\pi/2) + (1/2))$. Find the steady-state response $y_{ss}(n)$.

SOLUTION

To find the steady-state response, we need to use the formula in Equation (3.15). Thus, we need values for the parameters in the formula. The magnitude of the input is $X = 10$. M is the magnitude of $H(e^{j\theta}) = 1 + e^{-j\theta}$ as shown in previous examples. The input frequency is $\theta = \pi/2$, and $H(e^{j\theta})$ is

$$H\left(e^{j\pi/2}\right) = 1 + e^{-j\pi/2} = 1 + \cos\left(\frac{\pi}{2}\right) - j\,\sin\left(\frac{\pi}{2}\right) = 1 - j$$

The magnitude and the phase of the system at the frequency of the input are

$$M = \sqrt{1+1} = \sqrt{2}$$

$$\alpha = \tan^{-1}\left(\frac{-1}{1}\right) = -45° = -\frac{\pi}{4}$$

The steady-state response is then given by

$$y_{ss}(n) = 10\left(\sqrt{2}\right)\cos\left(\frac{\pi}{2}n + \frac{1}{2} - \frac{\pi}{4}\right) \quad -\infty \le n \le +\infty$$

3.5   FREQUENCY RESPONSE OF DISCRETE SYSTEMS

The frequency response of a discrete system is a set of output values when each value is obtained for a particular input $x(n)$ at a unique frequency value. At each frequency of the input, the output will have a change in its magnitude and phase. The frequency response $H(e^{j\theta})$ was derived in the process of finding the steady-state response in which we have found that

$$y_{ss}(n) = x(n)H\left(e^{j\theta n}\right)$$

if $x(n)$ is $e^{j\theta n}$ and we have defined the frequency response as

$$H\left(e^{j\theta}\right) = \sum_{m=-\infty}^{+\infty} h(m)e^{-j\theta m} \tag{3.16}$$

So if we have the impulse response of the system, we can use the previous equation to find the frequency response. Also we can calculate the frequency response $H(e^{j\theta})$ as a function of θ from the difference equation directly.

Consider the following first-order difference equation:

$$y(n) + ay(n-1) = x(n)$$

Let the input be periodic of the form

$$x(n) = e^{j\theta n}$$

We have seen that the output will be

$$y_{ss}(n) = e^{j\theta n} H(e^{j\theta})$$

Therefore, if we substitute this output in the difference equation, we will get

$$e^{j\theta n} H(e^{j\theta}) + ae^{j\theta(n-1)} H(e^{j\theta}) = e^{j\theta n}$$

If we cancel out the $e^{j\theta n}$ by dividing by $e^{j\theta n}$ (note that $e^{j\theta n}$ is never zero), we will have

$$H(e^{j\theta})(1 + ae^{-j\theta}) = 1$$

or

$$H(e^{j\theta}) = \frac{1}{1 + ae^{-j\theta}}$$

which is the frequency response of the system.

We can now generalize to the general difference equation case.

$$y(n) - \sum_{k=1}^{N} a_k y(n-k) = \sum_{k=0}^{M} b_k x(n-k) \tag{3.17}$$

Similar to what we did before and by letting $x(n) = e^{j\theta n}$ and $y(n) = e^{j\theta n} H(e^{j\theta})$, we will get

$$e^{j\theta n} H(e^{j\theta}) - \sum_{k=1}^{N} a_k e^{j\theta(n-k)} H(e^{j\theta}) = \sum_{k=0}^{M} b_k e^{j\theta(n-k)}$$

We can factorize $H(e^{j\theta})$ and cancel the $e^{j\theta n}$ to get

$$H(e^{j\theta})\left[1 - \sum_{k=1}^{N} a_k e^{-j\theta k}\right] = \sum_{k=0}^{M} b_k e^{-j\theta k}$$

and finally the general frequency response is

$$H(e^{j\theta}) = \frac{\displaystyle\sum_{k=0}^{M} b_k e^{-j\theta k}}{1 - \displaystyle\sum_{k=1}^{N} a_k e^{-j\theta k}} \tag{3.18}$$

Example 3.5

Find the frequency response for the following system:

$$y(n) + .5y(n-1) + 6y(n-2) = x(n) + x(n-1)$$

SOLUTION

With $x(n) = e^{j\theta n}$ and $y(n) = e^{j\theta n} H(e^{j\theta})$, the difference equation can be written as

$$e^{j\theta n} H\left(e^{j\theta}\right) + .5e^{j\theta(n-1)} H\left(e^{j\theta}\right) + 6e^{j\theta(n-2)} H\left(e^{j\theta}\right) = e^{j\theta n} + e^{j\theta(n-1)}$$

or

$$H\left(e^{j\theta}\right)\left[1 + .5e^{-j\theta} + .6e^{-j2\theta}\right] = 1 + e^{-j\theta}$$

The frequency response is then

$$H\left(e^{j\theta}\right) = \frac{1 + e^{-j\theta}}{1 + .5e^{-j\theta} + 6e^{-j2\theta}} = \frac{e^{j2\theta} + e^{j\theta}}{e^{j2\theta} + .5e^{j\theta} + 6}$$

3.5.1 PROPERTIES OF THE FREQUENCY RESPONSE

We will discuss two properties here.

3.5.1.1 Periodicity Property

We have seen the frequency response function $H(e^{j\theta})$ previously and it is presented here as

$$H\left(e^{j\theta}\right) = \sum_{m=-\infty}^{+\infty} h(m) e^{-j\theta m}$$

For periodicity, $H(e^{j\theta})$ must be the same as $H(e^{j(\theta+2\pi)})$.

$$H\left(e^{j(\theta+2\pi)}\right) = \sum_{m=-\infty}^{+\infty} h(m) e^{-j(\theta+2\pi)m} = \sum_{m=-\infty}^{+\infty} h(m) e^{-j\theta m} e^{-j2\pi m}$$

You can easily see that

$$e^{-j2\pi m} = \cos(2\pi m) - j \sin(2\pi m) = 1 - j(0) = 1$$

Thus,

$$H\left(e^{j(\theta+2\pi)}\right) = \sum_{m=-\infty}^{+\infty} h(m) e^{-j\theta m} = H\left(e^{j\theta}\right)$$

This property is important since in one period we can learn all about the signal that is being transformed.

3.5.1.2 Symmetry Property

To study this property, we will consider $H(e^{j\theta})$ for real values. We have

$$H\left(e^{j\theta}\right) = \sum_{m=-\infty}^{+\infty} h(m)e^{-j\theta m}$$

and

$$H\left(e^{-j\theta}\right) = \sum_{m=-\infty}^{+\infty} h(m)e^{j\theta m}$$

The magnitude of $e^{-j\theta m}$ is the same as the magnitude of $e^{j\theta n}$; thus,

$$\left| H\left(e^{j\theta}\right) \right| = \left| H\left(e^{-j\theta}\right) \right|$$

This indicates that the magnitude of $H(e^{j\theta})$ has even symmetry. Also the phase of $e^{-j\theta m}$ is $-\theta m$ and that of $e^{j\theta m}$ is θm. Therefore, the phase of $H(e^{j\theta})$ is the negative of the phase of $H(e^{-j\theta})$. This indicates that the phase of $H(e^{j\theta})$ has odd symmetry.

In general, from the frequency response for a given discrete system, you can tell if the system is able to pass a certain frequency range or reject another.

Example 3.6

Let

$$H\left(e^{j\theta}\right) = 1 - e^{-j\theta}$$

With respect to passing or rejecting frequencies, what kind of system is this?

SOLUTION

$H(e^{j\theta}) = 1 - e^{-j\theta} = 1 - (\cos\theta - j\sin\theta) = 1 - \cos\theta + j\sin\theta$. The magnitude response is

$$\left| H\left(e^{j\theta}\right) \right| = \sqrt{\left(\sin^2\theta\right) + \left(1 - \cos\theta\right)^2}$$

The plot of the magnitude of $H(e^{j\theta})$ versus θ is shown in Figure 3.5. Notice that the system attenuates low-frequency signals, but at frequencies approximately between $\pi/2$ and π this system does not attenuate as it does at frequencies near zero. Hence, this system passes high frequencies and prevents low frequencies from passing.

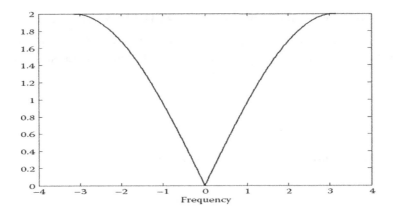

FIGURE 3.5   Magnitude plot for Example 3.6.

Example 3.7

Consider the following system with the frequency response function:

$$H\left(e^{j\theta}\right) = \frac{1}{2 - 0.1e^{-j\theta}}$$

What frequencies does this system pass and what frequencies does it not?

SOLUTION

The magnitude of the frequency response is

$$\left|H\left(e^{j\theta}\right)\right| = \left|\frac{1}{2 - .1e^{-j\theta}}\right| = \left|\frac{1}{2 - .1\cos\theta + .1j\,\sin\theta}\right| = \frac{1}{\sqrt{\left(2 - .1\cos\theta\right)^2 + \left(.1\sin\theta\right)^2}}$$

The plot of $|H(e^{j\theta})|$ is shown in Figure 3.6. It is clear from the plot that this system passes low frequencies and rejects high frequencies.

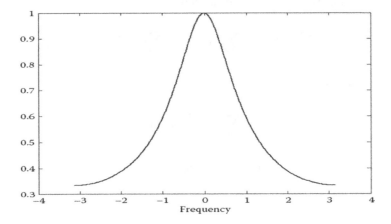

FIGURE 3.6   Magnitude plot for Example 3.7.

3.6 FOURIER TRANSFORM OF DISCRETE SIGNALS

The Fourier transform of a discrete signal $x(n)$ is $X(\theta)$, where $\theta = wT_s$, w is the analogue frequency of the continuous signal and T_s is the sampling period. The Fourier transform of discrete signals is defined as

$$X(\theta) = \sum_{n=-\infty}^{+\infty} x(n) e^{-j\theta n} \tag{3.19}$$

and to find $x(n)$ given $X(\theta)$, we use the following relation:

$$x(n) = \frac{1}{2\pi} \int_{2\pi} X(\theta) e^{j\theta n} d\theta \tag{3.20}$$

Note that $X(\theta)$ is a function of the continuous variable θ, the digital frequency, while $x(n)$ is a function of the discrete variable n.

We have seen that the frequency response function $H(e^{j\theta})$ was needed to calculate the steady-state response for a given discrete system. It also contains information about the system itself. The Fourier transform of discrete signals can be applied to signals that are periodic and not periodic. It will make the solution to difference equations much easier as we will see later.

Example 3.8

What is the Fourier transform of

$$x(n) = (.5)^n u(n)$$

SOLUTION

The Fourier transform of $x(n)$ is $X(\theta)$ and it is

$$X(\theta) = \sum_{n=-\infty}^{\infty} x(n) e^{-j\theta n} = \sum_{n=0}^{\infty} (.5)^n e^{-j\theta n} = \sum_{n=0}^{\infty} (.5 e^{-j\theta})^n = \left[\frac{1}{1 - .5 e^{-j\theta}} \right]$$

where we used the following geometric series sum:

$$\sum_{n=0}^{\infty} a^n = \frac{1}{1-a} \quad |a| < 1 \tag{3.21}$$

So $x(n) \leftrightarrow X(\theta)$ or we write the pairs $(.5)^n u(n) \leftrightarrow 1/(1 - .5 e^{-j\theta})$.

Example 3.9

What is the Fourier transform of the impulse signal $A\delta(n)$?

SOLUTION

For $x(n) = A\delta(n)$, we have

$$X(\theta) = \sum_{-\infty}^{\infty} A\delta(n)e^{-j\theta n} = A\delta(0)e^{-j\theta(0)} = A$$

Note that $\delta(n)$ is valid only at $n = 0$. Therefore,

$$x(n) = A\delta(n) \leftrightarrow X(\theta) = A$$

3.7 CONVERGENCE CONDITIONS

The Fourier transform of $x(n)$ is given again as

$$X(\theta) = \sum_{n=-\infty}^{+\infty} x(n)e^{-j\theta n}$$

where $X(\theta)$ is an infinite series and that series must converge for $X(\theta)$ to exist. For $X(\theta)$ to exist, it is necessary that $x(n)$ be summable. This means that

$$\sum_{n=-\infty}^{+\infty} |x(n)| < \infty$$

However, some signals such as the step signal are not summable. We will see how to deal with such signals later in this chapter.

3.8 PROPERTIES OF THE FOURIER TRANSFORM OF DISCRETE SIGNALS

Table 3.2 lists some properties of the Fourier transform of discrete signals. We will prove some of them here.

3.8.1 PERIODICITY PROPERTY

The Fourier transform of the discrete signal $x(n)$ is

$$X(\theta) = \sum_{n=-\infty}^{+\infty} x(n)e^{-j\theta n}$$

Let $\theta = \theta + 2\pi$. Then,

$$X(\theta + 2\pi) = \sum_{n=-\infty}^{+\infty} x(n)e^{-j\theta n}e^{-j2\pi n}$$

But $e^{-j2n\pi} = \cos(2\pi n) - j\sin(2\pi n) = 1 - 0 = 1$. Thus,

$$X(\theta + 2\pi) = \sum_{n=-\infty}^{+\infty} x(n)e^{-j\theta n} = X(\theta)$$

and $X(\theta)$ is periodic in θ with the period of 2π.

3.8.2 LINEARITY PROPERTY

The Fourier transform of $x(n) = a_1 x_1(n) + a_2 x_2(n)$ is

$$X(\theta) = \sum_{n=-\infty}^{n=\infty} \left[a_1 x_1(n) = a_2 x_2(n) \right] e^{-j\theta n}$$

$$X(\theta) = \sum_{n=-\infty}^{\infty} a_1 x_1(n)e^{-j\theta n} + \sum_{n=-\infty}^{\infty} a_2 x_2(n)e^{-j\theta n}$$

which clearly results in

$$X(\theta) = a_1 X_1(\theta) + a_2 X_2(\theta)$$

3.8.3 DISCRETE-TIME-SHIFTING PROPERTY

If $x(n)$ is shifted by n_0, then the Fourier transform of $x(n - n_0)$ is

$$\sum_{n=-\infty}^{+\infty} x(n - n_0)e^{-j\theta n}$$

If we let $n - n_0 = m$, then the Fourier transform of $x(n - n_0)$ becomes

$$\sum_{m=-\infty}^{+\infty} x(m)e^{-j\theta(m+n_0)} = e^{-j\theta n_0} \sum_{m=-\infty}^{\infty} x(m)e^{-j\theta m} = e^{-j\theta n_0} X(\theta)$$

The Fourier pairs are then

$$x(n - n_0) \leftrightarrow e^{-j\theta n_0} X(\theta)$$

3.8.4 FREQUENCY-SHIFTING PROPERTY

The Fourier transform of $x(n)ej\theta_0{}^n$, where θ_0 is the frequency shift is

$$\sum_{n=-\infty}^{+\infty} \left[e^{j\theta_0 n} x(n)e^{-j\theta n} \right] = \sum_{n=-\infty}^{+\infty} x(n)e^{-j(\theta-\theta_0)n} = X(\theta - \theta_0)$$

Therefore,

$$e^{j\theta_0 n} x(n) \leftrightarrow X(\theta - \theta_0)$$

3.8.5 REFLECTION PROPERTY

Consider the reflected signal $x(-n)$ of $x(n)$. Its Fourier transform is

$$\sum_{n=-\infty}^{+\infty} x(-n) e^{-j\theta n}$$

Let $-n = m$. Then the transform becomes

$$\sum_{m=-\infty}^{+\infty} x(m) e^{j\theta m} = \sum_{m=-\infty}^{+\infty} x(m) e^{-jm(-\theta)} = X(-\theta)$$

Thus,

$$x(-n) \leftrightarrow X(-\theta)$$

3.8.6 CONVOLUTION PROPERTY

In real time, with an input $x(n)$ and a system transfer function $h(n)$, the output of the system is then $y(n)$ and is given by

$$y(n) = x(n) * h(n) = \sum_{m=-\infty}^{+\infty} x(m) h(n - m) \tag{3.22}$$

Sometimes this summation becomes very complex to evaluate. Now let us take the Fourier transform on both sides to get

$$Y(\theta) = f[x(n) * h(n)] = f\left[\sum_{m=-\infty}^{\infty} x(m) h(n - m) \right]$$

where f indicates "the Fourier transform of." But according to the defining equation of the Fourier transform, we have

$$f\left[\sum_{m=-\infty}^{\infty} x(m) h(n - m) \right] = \sum_{n=-\infty}^{+\infty} \left[\sum_{m=-\infty}^{\infty} x(m) h(n - m) \right] e^{-j\theta n}$$

$$= \sum_{m=-\infty}^{+\infty} x(m) \sum_{n=-\infty}^{+\infty} h(n - m) e^{-j\theta n}$$

where the last term is obtained using the interchange of summation property.

Now let $k = n - m$ to get

$$Y(\theta) = \sum_{m=-\infty}^{+\infty} x(m) \sum_{k=-\infty}^{\infty} h(k)e^{-j\theta(k+m)} = \sum_{m=-\infty}^{+\infty} x(m)e^{-j\theta m} \sum_{k=-\infty}^{+\infty} h(k)e^{-j\theta k}$$

Finally, the important relation results in the equation for the convolution in the frequency domain.

$$Y(\theta) = X(\theta)H(\theta) \tag{3.23}$$

This result means that convolution in real time is multiplication in the frequency domain, an easy-to-manipulate complex algebra. This result can be written as

$$x_1(n) * x_2(n) \leftrightarrow X_1(\theta)X_2(\theta) \tag{3.24}$$

Example 3.10

Consider the discrete system with $h(n) = \delta(n) + \delta(n-1)$ and an input $x(n) = (.5)^n u(n)$. Find the output $y(n)$ using the Fourier transform method.

SOLUTION

The transform of $h(n)$ using Table 3.1 and the linearity and the shifting properties in Table 3.2 is

$$H(\theta) = 1 + e^{-j\theta}$$

and the Fourier transform of $x(n)$ is

$$X(\theta) = \frac{1}{1 - .5e^{-j\theta}}$$

The output using the Fourier transform is then

$$Y(\theta) = X(\theta)H(\theta) = \left[1 + e^{-j\theta}\right]\left[\frac{1}{1 - .5e^{-j\theta}}\right] = \frac{1}{1 - .5e^{-j\theta}} + \frac{e^{-j\theta}}{1 - .5e^{-j\theta}}$$

We can bring the previous equation into the discrete domain. The first term is $(.5)^n u(n)$. We can use the shifting property in Table 3.2 for the second term and get $(.5)^{n-1}u(n-1)$. The output in the discrete domain is

$$y(n) = (.5)^u u(n) + (.5)^{n-1} u(n-1) \ n \geq 0$$

TABLE 3.1
Fourier Transform Pairs

$x(n)$	$x(\theta)$
$A\delta(n)$	A
$Au(n)$	$A\left[\dfrac{1}{1-e^{-j\theta}} + \displaystyle\sum_{n=-\infty}^{\infty} \pi\delta(\theta - 2\pi n)\right]$
A	$2\pi A \displaystyle\sum_{n=-\infty}^{+\infty} \delta(\theta - 2\pi n)$
$a^n u(n)\ \lvert a\rvert < 1$	$\dfrac{1}{1 - ae^{-j\theta}}$
$na^n u(n)\ \lvert a\rvert < 1$	$\dfrac{ae^{j\theta}}{\left(e^{j\theta} - a\right)^2}$
$e^{j\theta_0 n}$	$2\pi \displaystyle\sum_{n=-\infty}^{+\infty} \delta(\theta - \theta_0 - 2\pi n)$
$\cos(\theta_0 n)$	$\pi \displaystyle\sum_{n=-\infty}^{+\infty} \left[\delta(\theta - \theta_0 - 2\pi n) + \delta(\theta + \theta_0 - 2\pi n)\right]$
$\sin(\theta_0 n)$	$\dfrac{\pi}{j} \displaystyle\sum_{n=-\infty}^{+\infty} \delta(\theta - \theta_0 - 2\pi n) - \delta(\theta + \theta_0 - 2\pi n)$

TABLE 3.2
Fourier Transform Properties

Real Time	Fourier Domain
$a_1 x_1(n) + a_2 x_2(n)$	$a_1 X_1(\theta) + a_2 X_2(\theta)$
$x(n - n_0)$	$e^{-j\theta_0 n} X(\theta)$
$e^{j\theta_0 n} x(n)$	$X(\theta - \theta_0)$
$nx(n)$	$j\dfrac{d}{d\theta}\left[X(\theta)\right]$
$x_1(n) * x_2(n)$	$X_1(\theta)X_2(\theta)$
$x_1(n)x_2(n)$	$\dfrac{1}{2\pi}\displaystyle\int_{-\infty}^{+\infty} X_1(m)X_2(\theta - m)\,dm$
$x(-n)$	$X(-\theta)$
$\displaystyle\sum_{n=-\infty}^{\infty} \lvert x(n)\rvert^2$	$\dfrac{1}{2\pi}\displaystyle\int_{-\pi}^{\pi} \lvert X(\theta)\rvert^2\,d\theta$

Example 3.11

Let us pass the input $x(n) = (.5)^u u(n)$ through two systems with

$$h_1(n) = h_2(n) = \delta(n) + \delta(n-1)$$

Find $y(n)$ using the Fourier transform method.

SOLUTION

As in Example 3.10,

$$X(\theta) = \frac{1}{1 - .5e^{-j\theta}}$$

and

$$H_1(\theta) = H_2(\theta) = 1 + e^{-j\theta}$$

If we call the output of the first system $y_1(n)$, then

$$Y_1(\theta) = X(\theta)H_1(\theta) = \left[\frac{1}{1 - .5e^{-j\theta}}\right]\left[1 + e^{-j\theta}\right] = \frac{1}{1 - .5e^{-j\theta}} + \frac{e^{-j\theta}}{1 - .5e^{-j\theta}}$$

The output of the second system will have the output of the first system as its input. Therefore, the output of the second system $y(n)$ is

$$Y(\theta) = Y_1(\theta)H_2(\theta) = \left[\frac{1}{1 - 0.5e^{-j\theta}} + \frac{e^{-j\theta}}{1 - .5e^{-j\theta}}\right]\left[1 + e^{-j\theta}\right]$$

$$= \frac{1}{1 - 0.5e^{-j\theta}} + \frac{e^{-j\theta}}{1 - 0.5e^{-j\theta}} + \frac{e^{-j\theta}}{1 - .5e^{-j\theta}} + \frac{e^{-2j\theta}}{1 - .5e^{-j\theta}}$$

By using Tables 3.1 and 3.2, we get

$$y(n) = (.5)^n u(n) + 2(.5)^{n-1} u(n-1) + (.5)^{n-2} u(n-2) \quad n \geq 0$$

3.9 PARSEVAL'S RELATION AND ENERGY CALCULATIONS

We have seen in Chapter 1 that the total energy in the signal $x(n)$ is

$$E = \sum_{n=-\infty}^{+\infty} |x(n)|^2$$

The Parseval's relation for discrete signals says that this energy can be computed using the Fourier transform as

$$E = \sum_{n=-\infty}^{+\infty} |x(n)|^2 = \frac{1}{2\pi} \int_{-\pi}^{\pi} |X(\theta)|^2 \, d\theta$$

Example 3.12

Find the energy in the signal.

$$x(n) = \delta(n) + \delta(n-1)$$

SOLUTION

We have seen that

$$E = \sum_{n=-\infty}^{+\infty} |x(n)|^2 = (1)^2 + (1)^2 = 2$$

We can use the Parseval's theorem to find the energy as

$$E = \frac{1}{2\pi} \int_{-\pi}^{\pi} \left| (1 + e^{-j\theta}) \right|^2 d\theta$$

$$E = \frac{1}{2\pi} \int_{-\pi}^{\pi} |(1 + \cos\theta - j\sin\theta)|^2 d\theta = \frac{1}{2\pi} \int_{-\pi}^{\pi} \left((1 + \cos\theta)^2 + \sin^2\theta \right) d\theta$$

If we simplify, we arrive at

$$E = \frac{1}{2\pi} \int_{-\pi}^{\pi} 1 d\theta + \frac{1}{2\pi} \int_{-\pi}^{\pi} 2\cos\theta \, d\theta + \frac{1}{2\pi} \int_{-\pi}^{\pi} \cos^2\theta \, d\theta + \frac{1}{2\pi} \int_{-\pi}^{\pi} \sin^2\theta \, d\theta$$

$$E = \frac{1}{2\pi}(\pi + \pi) + \frac{2}{2\pi}[\sin(\pi) - \sin(-\pi)] + \frac{1}{2\pi}(\pi + \pi) = 1 + 0 + 1 = 2$$

3.10 NUMERICAL EVALUATION OF THE FOURIER TRANSFORM OF DISCRETE SIGNALS

We should keep in mind that the Fourier transform of $x(n)$ is $X(\theta)$, which is a complex function in θ. MATLAB® can be used to calculate the complex values for X at each θ value. The magnitude of X, the phase of X, the real part of X, and the imaginary part of X can be calculated and plotted. We also know that $X(\theta)$ is periodic of period 2π. Since $X(\theta)$ is symmetric when $x(n)$ is real, we can plot $|X(\theta)|$ in the interval $[0, \pi]$ and know all about $x(n)$ in that range.

Example 3.13

Find the Fourier transform of $x(n) = (.1)^n$ for $n \geq 0$. Plot the magnitude and the phase of $X(\theta)$, the Fourier transform of $x(n)$.

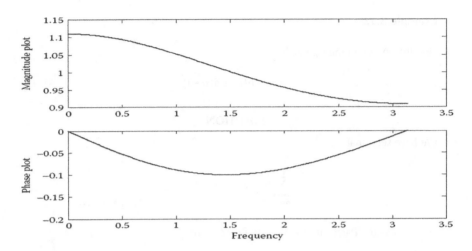

FIGURE 3.7   Frequency response for Example 3.13.

SOLUTION

The Fourier transform is given by

$$X(\theta) = \sum_{m=-\infty}^{+\infty} x(m)e^{-j\theta m} = \sum_{m=0}^{\infty}(.1)^m e^{-j\theta m} = \sum_{m=0}^{\infty}\left(0.1e^{-j\theta}\right)^m = \frac{1}{1-0.1e^{-j\theta}}$$

To use MATLAB to plot $|X(\theta)|$ and the phase of $X(\theta)$ versus θ we need to put $X(\theta)$ in the form

$$X(\theta) = \frac{a_0 + a_1 e^{-j\theta} + a_2 e^{-2j\theta} + a_3 e^{-3j\theta} + \cdots}{1 + b_1 e^{-j\theta} + b_2 e^{-2j\theta} + b_3 e^{-3j\theta} + \cdots} \tag{3.25}$$

We then can use the MATLAB function
  X = freqz (n, d, df)
The function freqz receives the numerator vector n, the denominator vector d, and the digital frequency vector, and sends back the complex values of $X(\theta)$ in the vector X. Then we can use the MATLAB functions abs and angle to find the magnitude and the angle of the frequency response. In our example,

$$X(\theta) = \frac{1}{1-.1e^{-j\theta}}$$

where $n = [1]$; $d = [1 - .1]$. We will use 401 frequency points in the range from 0 to π radians. The MATLAB script Example 313 will be used.
  The plots are shown in Figure 3.7.

Example 3.14

Find the Fourier transform of $x(n) = \delta(n) + \delta(n-1) + \delta(n-2) + \delta(n-3)$. Plot the magnitude and the phase of $X(\theta)$.

SOLUTION

Since $x(n)$ is finite and defined only at the $n = 0, 1, 2$, and 3, we can still use the MATLAB function freqz and we can use the Example3141 script to plot the magnitude and phase of $X(\theta)$ at 401 frequency points.

The plots are shown in Figure 3.8.

However, we can take a different approach and implement the defining equation for $X(\theta)$ directly. We have

$$X(\theta) = \sum_{m=-\infty}^{+\infty} x(m) e^{-j\theta m}$$

Assume that $x(n)$ is valid for $n_1 \leq n \leq n_2$ and we want to calculate $X(\theta)$ at $p + 1$ points. We are also interested in the frequency range from 0 to π. In this case, the frequency spacing is taken as

$$df = \left(\frac{\pi}{p}\right)m \quad \text{for } m = 0, 1, 2, \cdots, p$$

The frequency response becomes

$$X(m) = \sum_{r=1}^{N} e^{-j(\pi/p)mn_r} x(n_r) \quad \text{for } m = 0, 1, \cdots, p \tag{3.26}$$

where N is the number of samples for $x(n)$. Note that if $x(n_r)$ is a column vector having N rows and 1 column, and $X(m)$ is also a column vector having $(p + 1)$ rows and 1 column, then the summation

$$\sum_{r=1}^{N} e^{-j(\pi/p)mn_r}$$

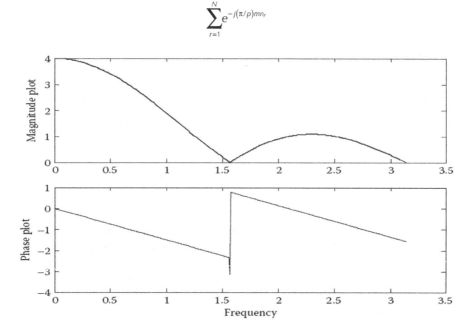

FIGURE 3.8 Frequency response for Example 3.14 using freqz MATLAB function.

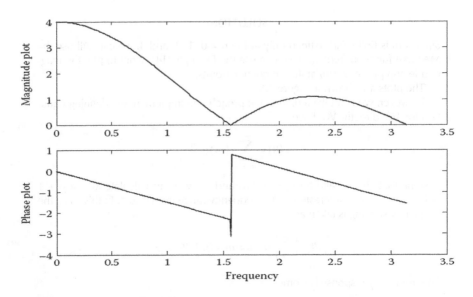

FIGURE 3.9   Frequency response for Example 3.14 using Equation (3.19) directly.

should be a matrix of $p + 1$ rows and N columns. This requires the vector m to be a column vector with $p + 1$ rows and 1 column and the n_r vector to be a row vector with 1 row and N columns. In MATLAB, if we enter data as row vectors then Equation (3.26) can be written as

$$X = e^{-j(\pi/p)m'n} x'$$ (3.27)

where ' indicates transpose of a matrix or a vector in MATLAB. To implement this method for the example at hand, we will write the script Example3142, and the plots are shown in Figure 3.9. Note that this example was straightforward. The method described in the example will be very useful, especially if you are to find the Fourier transform of $x(n) = n(.1)^n \cos(n)$ for $0 \le n \le 100$ or for $0 \le n \le 1000$ or for a bigger range. We will look at an example of this form in the Section 3.12.

3.11   SOME INSIGHTS: WHY IS THIS FOURIER TRANSFORM?

3.11.1   EASE IN ANALYSIS AND DESIGN

With the Fourier transform, we are able to identify the frequency contents of the input signal $x(n)$, both the magnitude and the phase spectrum. It may not be possible to predict what frequencies such $x(n)$ contains, especially if they are given as a plot. Knowing the frequency contents of the input signal gives us more insights into the way we analyze and design linear systems. We will know what frequencies will pass and what frequencies will not pass through a particular system. Some convolution sums are very difficult to evaluate in discrete real time. With the help of the Fourier transform, things become much easier.

3.11.2 SINUSOIDAL ANALYSIS

If the system $H(\theta)$ is subject to a sinusoidal input signal

$$x(n) = X\cos(\theta t + \varphi)$$

the steady-state output, $y_{ss}(n)$, of the system can be evaluated as

$$y_{ss}(n) = |x(n)||H(\theta)|\cos(\theta n + \varphi + \alpha)$$

or

$$y_{ss}(n) = XM\cos(\theta n + \varphi + \alpha)$$

where

X is the magnitude of the input $x(n)$
M is the magnitude of the frequency response of the discrete system $H(\theta)$
φ is the phase of $H(\theta)$ at a particular given θ
α is the phase of $H(\theta)$

This says that the output of a linear time-invariant system, if subject to a sinusoidal input, will have a steady-state solution equal to the magnitude of the input signal multiplied by the magnitude of the transfer function of the system evaluated at the frequency of the input signal and shifted by the phase angle of the transfer function evaluated at the input frequency as well.

Note also that the frequency of the output is the same as the frequency of the input. This means that the system is linear. If the system is not linear, the output frequency will *differ* from the input frequency.

3.12 END-OF-CHAPTER EXAMPLES

EOCE 3.1

Consider the discrete system given as follows:

$$h(n) = .1\delta(n) + .2\delta(n-2) + .5\delta(n-3)$$

Plot the frequency response, the phase and magnitude of $H(\theta)$.

SOLUTION

The impulse response $h(n)$ is defined only at $n = 0$, $n = 2$, and $n = 3$. With

$$H(e^{j\theta}) = \sum_{m=-\infty}^{+\infty} h(m)e^{-j\theta m}$$

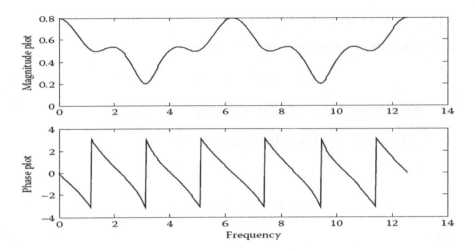

FIGURE 3.10 Frequency response for EOCE 3.1.

we have

$$H\left(e^{j\theta}\right) = .1e^{-j\theta(0)} + .2e^{-j2\theta} + .5e^{-j3\theta} = .1 + .2e^{-2j\theta} + .5e^{-3j\theta}$$

The plot of the phase and magnitude for $H(e^{j\theta})$ can be obtained using MATLAB and we will take the range from 0 to 4π to show that $H(e^{j\theta})$ is periodic in 0 of period 2π. The script is EOCE3_1

The plot is shown in Figure 3.10.

EOCE 3.2

Consider the following difference equation:

$$y(n) + .1y(n-1) + .2y(n-2) = x(n)$$

Find the frequency response and plot its magnitude and phase.

SOLUTION

With $x(n) = e^{j\theta n}$ and $y(n) = e^{j\theta n} H(e^{j\theta})$ we have from the given equation that

$$e^{j\theta n}H\left(e^{j\theta}\right) + .1e^{j\theta(n-1)}H\left(e^{j\theta}\right) + .2e^{j\theta(n-2)} = e^{j\theta n}$$

When we factor out $H(e^{j\theta})$, we will get

$$H\left(e^{j\theta}\right)\left(1 + .1e^{-j\theta} + .2e^{-2j\theta}\right) = 1$$

and

$$H\left(e^{j\theta}\right) = \frac{1}{1 + .1e^{-j\theta} + .2e^{-2j\theta}}$$

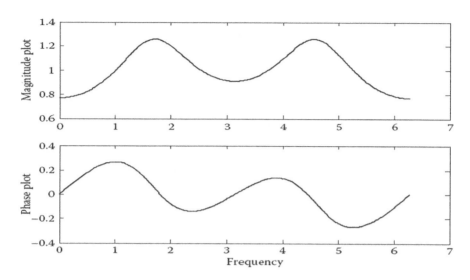

FIGURE 3.11 Frequency response for EOCE 3.2.

Next we use MATLAB to plot the magnitude and the phase of this frequency response in the range from 0 to 2π radians using EOCE3_2.
    The plot is shown in Figure 3.11.

EOCE 3.3

Find the frequency response of the system with

$$h(n) = n(.1)^n \sin(n) \quad \text{for } 0 \le n \le 100$$

SOLUTION

We know that

$$H\left(e^{j\theta}\right) = \sum_{m=-\infty}^{+\infty} h(m)e^{-j\theta m}$$

and this is not a difference equation, so we can use the MATLAB function freqz to find the frequency response. It is also not easy to find $H(e^{j\theta})$ analytically. So we use the method we established earlier in the chapter. In this case n_1 is 0 and n_2 is 100. We will take 400 points, so p is 400. The frequency resolution is taken as

$$df = \frac{\pi}{p}m \quad m = 0, 1, 2, \cdots, p$$

We will use MATLAB to implement $H = e^{-j(\pi/p)m,n}h'$ as we did earlier for the Fourier transform approximation. The MATLAB script used is EOCE3_3.
    The plots are shown in Figure 3.12.

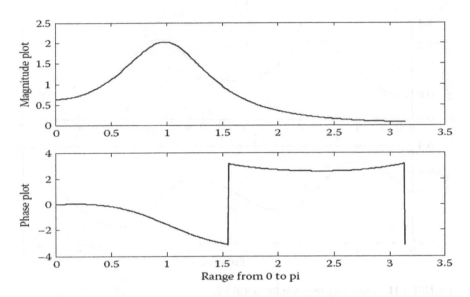

FIGURE 3.12   Frequency response for EOCE 3.3.

EOCE 3.4

Consider the discrete system described by the impulse response.

$$h(n) = (.9)^n u(n)$$

Find that steady-state response of the system when

$$x(n) = 2\cos\left(\frac{\pi}{2}n\right) + 4\sin\left(\frac{\pi}{2}n\right)$$

SOLUTION

First we need to find $H(e^{j\theta})$, and then we can use the formula

$$y_{ss}(n) = \left|H\left(e^{j\theta}\right)\right|\left|x(n)\right|\cos(\theta n + \phi + \alpha)$$

to find the steady-state response. The frequency response is

$$H\left(e^{j\theta}\right) = \sum_{m=0}^{\infty}(.9)^m e^{-j\theta m} = \sum_{m=0}^{\infty}\left((.9)e^{-j\theta}\right)^m = \frac{1}{1-.9e^{-j\theta}} = \frac{1}{1-.9\cos\theta+.9\sin\theta}$$

The input can be divided into two separate inputs, $x_1(n)$ and $x_2(n)$, where

$$x_1(n) = 2\cos\left(\frac{\pi}{2}n\right) \quad \text{and} \quad x_2(n) = 4\sin\left(\frac{\pi}{2}n\right)$$

Then we can use superposition to find $y_{ss}(n) = y_{ss1}(n) + y_{ss2}(n)$, where $y_{ss1}(n)$ is the output due to $x_1(n)$ and $y_{ss2}(n)$ is the output due to $x_2(n)$.

For $x_1(n) = 2\cos(\pi/2n)$, we have

$$|x_1(n)| = 2, \quad \theta = \frac{\pi}{2} \quad \text{and} \quad \varphi = 0$$

With

$$|H(e^{j\theta})| = \frac{1}{\sqrt{(1-.9\cos\theta)^2 + (.9\sin\theta)^2}}$$

and for $\theta = \pi/2$, the magnitude of the frequency response becomes

$$|H(e^{j\pi/2})| = \frac{1}{\sqrt{(1-.9(0))^2 + (.9(1))^2}} = \frac{1}{\sqrt{1+.81}} = \frac{1}{\sqrt{2.81}}$$

and the phase is $0 - \tan^{-1}(.9/1) = 0.7328$. The steady-state solution is then

$$y_{ss1}(n) = 2\left(\frac{1}{\sqrt{2.81}}\right)\cos\left(\frac{\pi}{2}n + 0 - \tan^{-1}\left(\frac{.9}{1}\right)\right) - \infty \le n \le +\infty$$

For $x_2(n) = 4\sin(\pi/2)$, we need to write both inputs using the same reference. We write this second input as

$$x_2(n) = 4\cos\left(\frac{\pi}{2}n - \frac{\pi}{2}\right)$$

The only difference now is in the phase shift of the input and its magnitude.

$$|x_2(n)| = 4 \quad \text{and} \quad \varphi = -\frac{\pi}{2}$$

The output is then

$$y_{ss1}(n) = 4\left(\frac{1}{\sqrt{2.81}}\right)\cos\left(\frac{\pi}{2}n - \frac{\pi}{2} - \tan^{-1}\left(\frac{.9}{1}\right)\right) - \infty \le n \le +\infty$$

Finally,

$$y_{ss1}(n) = y_{ss1}(n) + y_{ss2}(n)$$

$$= \left(\frac{1}{\sqrt{2.81}}\right)\left[2\cos\left(\frac{\pi}{2}n + 0 - 0.7328\right) + 4\cos\left(\frac{\pi}{2}n - \frac{\pi}{2} - 0.7328\right)\right]$$

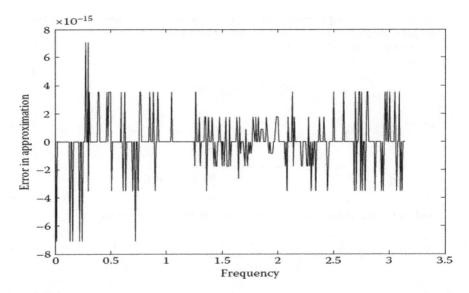

FIGURE 3.13   Error for EOCE 3.5.

EOCE 3.5

The Fourier transform of $\alpha x_1(n) + \beta x_2(n)$ is $\alpha X_1(\theta) + \beta X_2(\theta)$. Use MATLAB to show this property.

SOLUTION

Consider the following two signals:

$$x_1(n) = 2\delta(n) + 3\delta(n-1) - 4\delta(n-3)$$

and

$$x_2(n) = 2\delta(n) + 3\delta(n-3)$$

Let $\alpha = 2$ and $\beta = 3$. Now consider the MATLAB script EOCE3_5 to obtain the proof. The plot is shown in Figure 3.13.

EOCE 3.6

Use MATLAB to prove the time-shifting property of the Fourier transform of discrete signals.

SOLUTION

Consider the following signal:

$$x(n) = \delta(n) + 2\delta(n-1) + 3\delta(n-2)$$

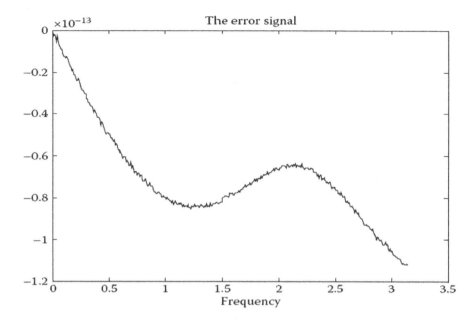

FIGURE 3.14   Error for EOCE 3.6.

We know that by shifting $x(n)$ three units, for example, only the index n for $x(n)$ will change. The MATLAB script is EOCE3_6.
    The plot is shown in Figure 3.14.

EOCE 3.7

Consider the following frequency response:

$$H\left(e^{j\theta}\right) = \frac{1}{1 - .9e^{-j\theta}}$$

1. What is the steady-state output if the input to the system is

$$x(n) = \frac{1}{2}\cos\left(\frac{\pi}{4}n\right)$$

2. Use MATLAB to find the response to this input.

SOLUTION

1. The frequency response at the input frequency is

$$H\left(e^{j\pi/4}\right) = \frac{1}{1 - .9\left(\sqrt{2}/2\right) + .9\left(\sqrt{2}/2\right)j}$$

and the magnitude at the input frequency is

$$\left|H\left(e^{j\pi/4}\right)\right| = \frac{1}{\sqrt{\left(1-.9\left(\sqrt{2}/2\right)\right)^2+\left(.9\left(\sqrt{2}/2\right)\right)^2}} = 1.36$$

The angle is

$$0-\tan^{-1}\left(\frac{\left(.9\sqrt{2}/2\right)}{1-.9\left(\sqrt{2}/2\right)}\right) = -1.05$$

and the steady-state solution is

$$y_{ss}(n) = \left(\frac{1}{2}\right)(1.36)\cos\left(\frac{\pi}{4}n+(-1.05)\right)$$

2. We can use MATLAB and write the script EOCE3_7.

The plots are shown in Figure 3.15. You can see that as n increases in the solution using MATLAB for the total response, the outputs in the steady-state solution and the MATLAB total response do match.

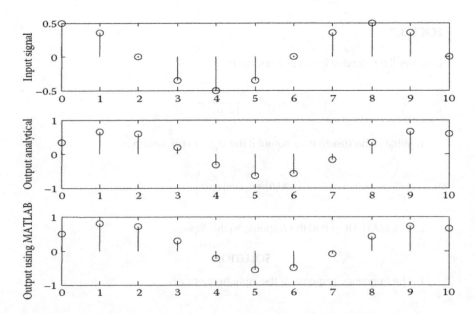

FIGURE 3.15   Signals for EOCE 3.7.

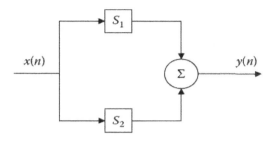

FIGURE 3.16 System for EOCE 3.8.

EOCE 3.8

Consider the system shown in Figure 3.16. Let S_1 be represented by

$$y_1(n) + .1y_1(n-1) = x(n)$$

and S_2 be represented by

$$y_2(n) + .1y_2(n-1) + 2y(n-2) = x(n)$$

Find the output using MATLAB if

$$x(n) = \frac{1}{2}\cos\left(\frac{\pi}{4}n - \pi\right)$$

SOLUTION

The frequency response of system 1 is obtained as

$$H_1(e^{j\theta}) = \frac{1}{1 + .1e^{-j\theta}}$$

and the frequency response of system 2 is obtained as

$$H_2(e^{j\theta}) = \frac{1}{1 + .1e^{-j\theta} + 2e^{-j2\theta}}$$

Systems 1 and 2 are connected in parallel and $y(n)$ can be calculated as the output of the whole system. Thus,

$$H_1(e^{j\theta}) + H_2(e^{j\theta}) = \frac{1 + .1e^{-j\theta} + 2e^{-j2\theta} + 1 + .1e^{-j\theta}}{\left(1 + .1e^{-j\theta}\right)\left(1 + .1e^{-j\theta} + 2e^{-j2\theta}\right)}$$

$$= \frac{2 + .2e^{-j\theta} + 2e^{-j2\theta}}{1 + .2e^{-j\theta} + 2.01e^{-j2\theta} + .2e^{-j3\theta}}$$

The MATLAB script used to find the output is EOCE3_8
  The plot is shown in Figure 3.17.

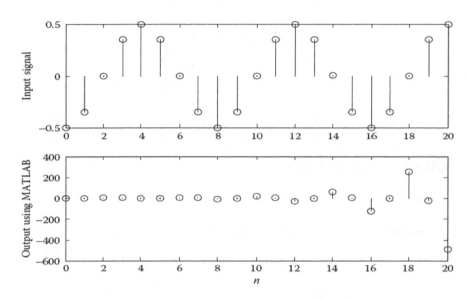

FIGURE 3.17 Signals for EOCE 3.8.

EOCE 3.9

Consider the system in Figure 3.18. System S_1 is represented by

$$y_1(n) + y_1(n-1) = x(n)$$

and S_2 is represented by

$$y_2(n) - y_2(n-1) = y_1(n)$$

Find the steady-state response $y(n)$ if $x(n)$ is

$$x(n) = .5\cos\left(n\frac{\pi}{4}\right)$$

Find also the response to this input using MATLAB. Note that using MATLAB you will get the total response. The steady-state response is obtained by observing the MATLAB output for large n.

FIGURE 3.18 System for EOCE 3.9.

SOLUTION

Since S_1 and S_2 are connected in series, the steady-state $y_{ss}(n)$ can be thought of as the output to the system $H_1(e^{j\theta})\, H_2(e^{j\theta})$ with the input $x(n)$.

$$H_1\!\left(e^{j\theta}\right) = \frac{1}{1+e^{-j\theta}} \quad \text{and} \quad H_2\!\left(e^{j\theta}\right) = \frac{1}{1-e^{-j\theta}}$$

The whole system is now

$$H\!\left(e^{j\theta}\right) = H_1\!\left(e^{j\theta}\right) H_2\!\left(e^{j\theta}\right) = \frac{1}{1+e^{-j\theta}}\,\frac{1}{1-e^{-j\theta}} = \frac{1}{1-e^{-2j\theta}} = \frac{1}{1-\cos 2\theta + j\sin 2\theta}$$

The frequency response at the input frequency is given by

$$H\!\left(e^{j\pi/4}\right) = \frac{1}{1-\cos\left(2\pi/4\right)+j\sin\left(2\pi/4\right)} = \frac{1}{1+j}$$

The magnitude of the frequency response is $1/\sqrt{2}$ and its phase is $(0 - \tan^{-1}(1/1)) = -\pi/4$. Thus, the steady-state response for the whole system is

$$y_{ss}(n) = \left(\frac{1}{2}\right)\left(\frac{1}{\sqrt{2}}\right)\cos\left(n\frac{\pi}{4}-\frac{\pi}{4}\right) = \frac{1}{2\sqrt{2}}\cos\left(\frac{\pi}{4}(n-1)\right)$$

We see that the output is the input reduced in magnitude by $1/\sqrt{2}$ and shifted by $-\pi/4$. We can use MATLAB to plot $y_{ss}(n)$ and the response to the given input. We write the script EOCE3_9.

The plots are shown in Figure 3.19.

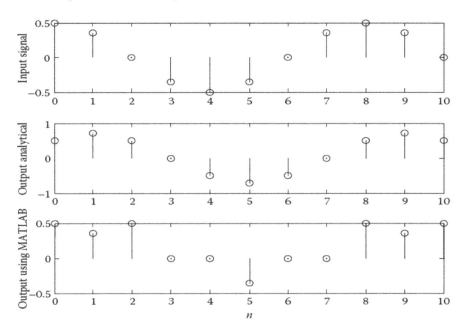

FIGURE 3.19   Plots for EOCE 3.9.

EOCE 3.10

Two systems $h_1(n)$ and $h_2(n)$ are connected in series with an input $x(n) = \delta(n)$. Find the output $y(n)$ for $n \geq 0$ if $h_1(n) = (1/2)^n u(n)$ and $= h_2(n) = (1/4)^n u(n)$.

SOLUTION

In real time $y(n) = (h_1(n) *h_2(n)) * x(n)$. But we can use the Fourier transform to write

$$Y(\theta) = X(\theta)H_1(\theta)H_2(\theta)$$

With $X(\theta) = 1$, the first system is

$$H_1(\theta) = \frac{1}{1-.5e^{-j\theta}}$$

and the second system is

$$H_2(\theta) = \frac{1}{1-(1/4)e^{-j\theta}}$$

The output then is given as

$$Y(\theta) = (1)\left(\frac{1}{1-.5e^{-j\theta}}\right)\left(\frac{1}{1-(1/4)e^{-j\theta}}\right)$$

$$= \frac{2}{1-(1/2)e^{-j\theta}} - \frac{1}{1-(1/4)e^{-j\theta}}$$

Using Table 3.1 we get

$$y(n) = 2\left(\frac{1}{2}\right)^n - \left(\frac{1}{4}\right)^n \quad n \geq 0$$

with

$$H(\theta) = \frac{1}{1-(3/4)e^{-j\theta} + (1/8)e^{-j2\theta}}$$

Let us plot $y(n)$ analytically and using MATLAB script EOCE3_10. The plots are shown in Figure 3.20.

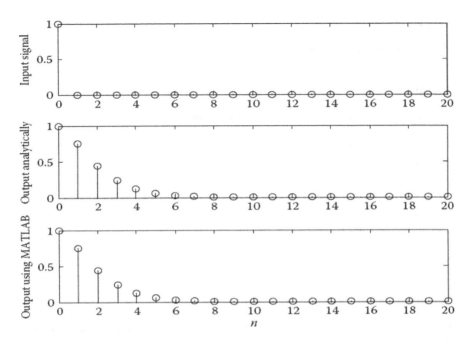

FIGURE 3.20 Plots for EOCE 3.10.

3.13 END-OF-CHAPTER PROBLEMS

EOCP 3.1

Perform the following calculations and give the result in the form $a + jb$:

1. $\dfrac{1-j}{j} - 1$

2. $\dfrac{1 - je^{-j\pi/2}}{1+j}$

3. $\dfrac{1}{j} - \dfrac{1}{j-1} + j$

4. $\dfrac{e^{\pi} - e^{-j\pi}}{j} + 1$

5. $e^{j\pi/2} + \dfrac{e^{j\pi} - 1}{1 - j}$

EOCP 3.2

Put the complex numbers in EOCP 3.1 in the polar form.

EOCP 3.3

Find the frequency response for each of the following systems:

1. $h(n) = \delta(n) + \delta(n-1)$
2. $h(n) = \delta(-n+2) + \delta(-n+1) + \delta(n) + \delta(n-1) + \delta(n-2)$
3. $h(n) = a(b)nu(n)$
4. $h(n) = a(b)^n u(n) - a(b)^{n-1} u(n-1)$
5. $h(n) = u(n-4)$
6. $h(n) = a(b)^n u(n) - a(b)^{n-4} u(n-4)$
7. $h(n) = (a)^n \cos(n\pi)u(n)$
8. $h(n) = (a)^n \cos(n\pi)u(n) - (a)^{n-1} \cos(n\pi - \pi)u(n-1)$

EOCP 3.4

For $a = .5$, and $b = .4$, use MATLAB to plot the magnitude and phase of the frequency responses found in EOCP 3.3.

EOCP 3.5

Use MATLAB to find the steady-state response of each system in EOCP 3.3 to the following inputs:

1. $x(n) = \cos\left(\dfrac{\pi}{2}n\right)u(n)$

2. $x(n) = \cos\left(\dfrac{\pi}{2}n\right)u(n) - \cos\left(\dfrac{\pi}{2}(n-1)\right)u(n-1)$

EOCP 3.6

Find the frequency response for each of the following systems:

1. $y(n) + ay(n-1) = x(n)$
2. $y(n) + ay(n-1) = x(n) + x(n-1)$
3. $y(n) + ay(n-1) + by(n-2) = x(n)$
4. $y(n) + ay(n-2) = x(n)$
5. $y(n) + ay(n-3) = x(n) + x(n-3)$
6. $y(n) - ay(n-2) = x(n-3)$
7. $y(n) - ay(n-3) - by(n-4) = x(n)$

8. $y(n) - ay(n - 6) = x(n - 3)$
9. $y(n) + ay(n - 1) = x(n) + x(n - 1) + x(n - 3)$
10. $y(n) = x(n) + x(n - 1) + x(n - 3) + x(n - 4)$

EOCP 3.7

Use MATLAB to plot the magnitude and phase of the systems in EOCP 3.6 for $a = 1$ and $b = 5$.

EOCP 3.8

Use MATLAB to find the steady-state responses for the system in EOCP 3.6 with a and b as given in EOCE 3.7 if

1. $x(n) = 10$

2. $x(n) = \sin\left(\dfrac{2\pi}{3}n\right)u(n)$

EOCP 3.9

Find the Fourier transform of the following signals:

1. $3\delta(n) + 3\delta(n - 1)$
2. $(.5)^n u(n) + e^{jn}$
3. $(.5)^n u(n - 1) + e^{j(n-1)}$
4. $(.5)^n u(n) - (.5)^{n-1} u(n - 1)$
5. $(n - 1)(.5)^{n-1} u(n - 1)$
6. $3\delta(n) * 3\delta(n - 1)$
7. $(.5)^{n-1} u(n - 1) * e^{j(n-1)} u(n - 1)$
8. $(.5)^n u(n) * 3\delta(n) * 3\delta(n - 1)$
9. $(.5)^{n-1} u(n - 1) * \delta(n) * \delta(n)$
10. $(.5)^n u(n) + 3\delta(n) * 3\delta(n - 1)$

EOCP 3.10

Use the Fourier transform to find the output $y(n)$ when the input is $x(n) = (.5)^n u(n)$ for the following systems:

1. $h(n) = \delta(n) + \delta(n - 1)$
2. $h(n) = \delta(n - 1) + \delta(n - 2)$
3. $h(n) = (.5)^n u(n)$
4. $h(n) = (.5)^n u(n) - (.5)^{n-1} u(n - 1)$
5. $h(n) = (.3)^{n-5} u(n - 5) + \delta(n)$

EOCP 3.11

Consider the following systems shown in Figures 3.21–3.25. Find $y(n)$ if $x(n) = (.5)^n u(n)$. Use the Fourier transform method. We are given that

$$h_1(n) = \delta(n) + \delta(n-1)$$

$$h_2(n) = (.5)^n u(n)$$

$$h_3(n) = (.3)^{n-5} u(n-5) + \delta(n) = h(n)$$

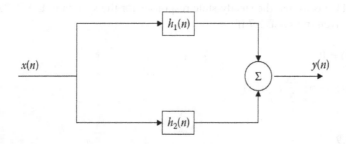

FIGURE 3.21   Signal for EOCP 3.11.

FIGURE 3.22   Signal for EOCP 3.11.

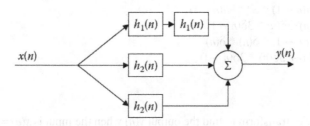

FIGURE 3.23   Signal for EOCP 3.11.

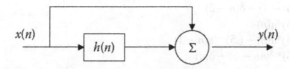

FIGURE 3.24   Signal for EOCP 3.11.

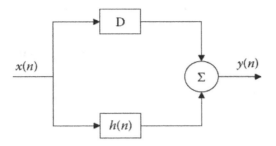

FIGURE 3.25 Signal for EOCP 3.11.

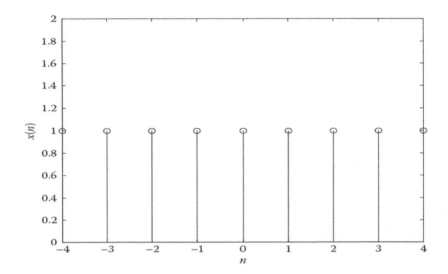

FIGURE 3.26 Signal for EOCP 3.12.

EOCP 3.12

Consider the signal shown in Figure 3.26. Find the Fourier transform of this signal.

EOCP 3.13

Consider the following system:

$$y(n) = x(n) - x(n-1)$$

1. Find the steady-state output if $x(n) = A \cos(n\theta)$.
2. Find the steady-state response when

$$x(n) = x_1(n) + x_2(n)$$

where

$$x_1(n) = B\cos(n\theta) \quad \text{and} \quad x_2(n) = C\cos(100n\theta)$$

3. What do you think that this system is doing?

EOCP 3.14

Consider the following system:

$$y(n) - y(n-1) = \frac{x(n) + x(n+1)}{2}$$

1. Find the frequency response of this system.
2. Find the steady-state response when

$$x(n) = \left(A\cos\left(\frac{n\pi}{3}\right)u(n) \right)$$

3. What do you say about this system?

EOCP 3.15

Consider the following systems:

$$H\left(e^{j\theta}\right) = 10 + 2e^{-j\theta} + 3e^{-3j\theta}$$

and

$$H\left(e^{j\theta}\right) = \frac{1}{1 + .5e^{-j\theta} - 5e^{-j2\theta}}$$

1. Find the difference equation that describes the two systems.
2. What is $h(n)$, the impulse response for both?
3. Use MATLAB to find the steady-state response if

$$x(n) = 10\cos\left(\frac{n\pi}{3}\right)u(n)$$

4. Are both systems stable?
5. Use MATLAB to plot the magnitude response of each system.
6. What kind of system is each?

EOCP 3.16

Consider the following discrete signals with $T_s = 0.1$ s:

1. $x(n) = \sin\left(\dfrac{4}{3}\pi n\right)$

2. $x(n) = \sin\left(\dfrac{8}{3}\pi n\right)$

3. $x(n) = \{1\ 1\ 1\}$ a periodic signal with $N = 3$.
   a. Find the period for the first two signals.
   b. Find the Fourier series coefficients.
   c. Where are these frequency components located?

4 z-Transform and Discrete Systems

4.1 INTRODUCTION

The z-transform is a frequency domain representation that makes solution, design, and analysis of discrete linear systems simpler. It also gives some insights about the frequency contents of signals where these insights are hard to see in real-time systems. There are other important uses for the z-transform, but we will concentrate only on the issues described in this introduction.

4.2 BILATERAL z-TRANSFORM

The z-transform of the signal $x(n)$ is given by

$$X(z) = \sum_{n=-\infty}^{+\infty} x(n)z^{-n} \qquad (4.1)$$

where z is the complex variable. If we try to expand Equation (4.1), we get

$$X(z) = \cdots + x(-2)z^2 + x(-1)\,z^1 + x(0)z^0 + x(1)z^{-1} + x(2)z^{-2} + \cdots \qquad (4.2)$$

You can see that in Equation (4.1), the power of z indicates the position of the samples in the signal $x(n)$. This notice is very important. Consider that

$$x(n) = x(0)\delta(n)$$

where this signal has the strength $x(0)$ and is located only at $n = 0$. The z-transform of $x(n)$, $X(z)$, is then

$$X(z) = \sum_{n=-\infty}^{+\infty} x(n)z^{-n} = \sum_{n=-\infty}^{+\infty} x(0)\delta(n)z^{-n} = x(0)z^0 = x(0)$$

Similarly, if

$$x(n) = x(-1)\delta(n+1) + x(0)\delta(n) + x(1)\delta(n-1)$$

then we can see that this signal has values only at $n = -1$, $n = 0$, and $n = 1$.

Thus,

$$X(z) = x(-1)z^1 + x(0)z^0 + x(1)z^{-1} = x(-1)z^1 + x(0) + x(1)z^{-1}$$

In general, if

$$x(n) = x(p)\delta(n - n_0)$$

is a signal that is available only at $n = n_0$, then $X(z)$ is

$$X(z) = x(p)(z)^{-n_0}$$

Example 4.1

Consider the signal in Figure 4.1. What is the z-transform of $x(n)$?

SOLUTION

$x(n)$ can be written as

$$x(n) = 2\delta(n+2) - 1\delta(n+1) + 2\delta(n) - 1\delta(n-1) + 2\delta(n-2)$$

and its z-transform is given by Equation (4.1) as

$$X(z) = \sum_{n=-\infty}^{+\infty} (x(n))z^{-n}$$

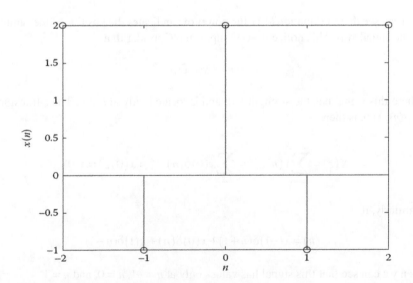

FIGURE 4.1 Signal for Example 4.1.

Substituting in the previous equation, we get

$$X(z) = 2z^2 - 1z^1 + 2z^0 - z^{-1} + 2z^{-2} = 2z^{-2} - z^1 + 2 - z^{-1} + 2z^{-2}$$

Notice that z^{-2} represents a delay of $2T_s$ units of time and z^3 represents an advance of $3T_s$ units of time, where T_s is the sampling interval for the signal $x(n)$.

4.3 UNILATERAL z-TRANSFORM

The unilateral z-transform is the transform of the signal $x(n)$ for $n \geq n_0$. We will take $n_0 = 0$ in this discussion. We will use the notation

$$x(n) \leftrightarrow X(z)$$

to indicate that we can get $X(z)$ from $x(n)$ and we can get $x(n)$ from $X(z)$ as well.

Example 4.2

Find the z-transform of $x(n) = A\delta(n)$.

SOLUTION

Using the definition of the z-transform, we write

$$X(z) = \sum_{n=0}^{+\infty} x(n) z^{-n} = \sum_{n=0}^{+\infty} A\delta(n) z^{-n}$$

But $\delta(n)$ is defined only at $n = 0$. So $X(z) = Az^{-0} = A$ and we write

$$A\delta(n) \leftrightarrow A$$

Similarly, we have

$$A\delta(n - n_0) \leftrightarrow Az^{-n_0}$$

Example 4.3

Find the z-transform of $x(n) = Au(n)$, the unit step discrete signal.

SOLUTION

With Equation (4.1), we have

$$X(z) = \sum_{n=-\infty}^{+\infty} x(n) z^{-n}$$

Since $u(n)$ starts at $n = 0$ and is available only for $n \geq 0$, $X(z)$ becomes

$$X(z) = \sum_{n=0}^{\infty} Az^{-n} = A \sum_{n=0}^{\infty} \left(z^{-1}\right)^n = \frac{A}{1-z^{-1}}$$

where the last result is a direct application of the geometric series sum. Thus, we write

$$Au(n) \leftrightarrow \frac{A}{1-z^{-1}}$$

and similarly,

$$Au(n - n_0) \leftrightarrow \frac{Az^{-n_0}}{1-z^{-1}}$$

Example 4.4

Find the z-transform of the signal $x(n) = Aa^n$ for $n \geq 0$.

SOLUTION

Using the defining equation of the z-transform, we write using the unilateral case

$$X(z) = \sum_{n=0}^{+\infty} x(n)z^{-n} = \sum_{n=0}^{+\infty} Aa^n z^{-n} = A \sum_{n=0}^{\infty} (az^{-1})^n = \frac{A}{1-az^{-1}}$$

Therefore, we write

$$Aa^n u(n) \leftrightarrow \frac{A}{1-az^{-1}}$$

and

$$Aa^{n-p} u(n - p) \leftrightarrow \frac{Az^{-p}}{1-az^{-1}}$$

Example 4.5

Find the z-transform of the complex exponential discrete signal

$$x(n) = Aa^n e^{j\theta n} u(n)$$

SOLUTION

Using the defining equation again, we write

$$X(z) = \sum_{n=0}^{+\infty} Aa^n e^{j\theta n} u(n) z^{-n} = A \sum_{n=0}^{+\infty} \left(ae^{j\theta} z^{-1}\right)^n = \frac{A}{1-ae^{j\theta}z^{-1}}$$

Using the notation we established, we write

$$Aa^n e^{j\theta n} u(n) \leftrightarrow \frac{A}{1 - ae^{j\theta}z^{-1}}$$

Example 4.6

Find the z-transform of the signal

$$x(n) = Aa^n \cos(\theta n)u(n)$$

SOLUTION

Using the defining equation, we get

$$X(z) = \sum_{n=0}^{+\infty} x(n)z^{-1} = \sum_{n=0}^{+\infty} Aa^n \cos(\theta n)u(n)z^{-n} = \frac{A}{2}\sum_{n=0}^{\infty} a^n(e^{j\theta n} + e^{-j\theta n})z^{-n}$$

By rearranging terms and using the geometric series sum, we arrive at

$$X(z) = \frac{A}{2}\sum_{n=0}^{\infty}\left(ae^{j\theta}z^{-n}\right)^2 + \frac{A}{2}\sum_{n=0}^{\infty}\left(ae^{-j\theta}z^{-n}\right)^n = \frac{A}{2}\left[\frac{1}{1 - ae^{j\theta}z^{-1}} + \frac{1}{1 - ae^{-j\theta}z^{-1}}\right]$$

After simplification, we get

$$Aa^n \cos(\theta n)u(n) \leftrightarrow \frac{A(1 - az^{-1}\cos\theta)}{1 - az^{-1}\left(e^{j\theta} + e^{-j\theta}\right) + a^2 z^{-2}}$$

4.4 CONVERGENCE CONSIDERATIONS

The z-transform of $x(n)$ is given by

$$X(z) = \sum_{n=0}^{+\infty} x(n)z^{-n}$$

where z is a complex number. This complex number can be written in polar form as

$$z = re^{j\theta}$$

With z in polar form, $X(z)$ becomes

$$X\left(re^{j\theta}\right) = \sum_{n=0}^{+\infty} x(n)\left(re^{j\theta}\right)^{-n} = \sum_{n=0}^{+\infty} x(n)r^{-n}e^{-j\theta n} \tag{4.3}$$

So we can see that the z-transform of $x(n)$ is the Fourier transform of $x(n)r^{-n}$. If $r = 1$ then $z = e^{j\theta}$ and $|z| = |e^{j\theta}| = 1$, which is a circle of unity magnitude radius.

In Equation (4.3), and if we consider the unilateral case, the series must converge for the z-transform to exist. This will happen if $x(n)r^{-n}$ is absolutely summable. Mathematically, we require

$$\sum_{n=0}^{+\infty} \left| x(n)r^{-n} \right| < \infty$$

The region where the z-transform converges is called the region of convergence (ROC) and is usually an annular region.

Example 4.7

What is the ROC of the z-transform of $x(n) = Au(n)$?

SOLUTION

We have seen that

$$x(n) = Au(n) \leftrightarrow \frac{A}{1-z^{-1}} = \frac{Az}{z-1}$$

The ROC is $|z| > 1$, which is the region exterior to the unit circle in the z-plane.

Example 4.8

What is the ROC of the z-transform of $x(n) = Aa^n u(n)$?

SOLUTION

We have seen that

$$Aa^n u(n) \leftrightarrow \frac{A}{1-az^{-1}} = \frac{Az}{z-a}$$

The ROC is then $|z| > |a|$.

Example 4.9

What is the ROC of the z-transform of $x(n) = Aa^n e^{j\theta n} u(n)$?

SOLUTION

We have seen that

$$Aa^n e^{j\theta n} u(n) \leftrightarrow \frac{A}{1-ae^{j\theta}z^{-1}} = \frac{Az}{z-ae^{j\theta}}$$

for which the ROC is $|z| > |ae^{j\theta}| = |a|$.

Example 4.10

What is the ROC of the z-transform of $x(n) = Aa^n \cos(\theta n) u(n)$?

SOLUTION

We have seen that

$$Aa^n \cos(\theta n) a(n) \leftrightarrow \frac{A}{2}\left[\frac{1}{1 - ae^{j\theta}z^{-1}} + \frac{1}{1 - ae^{-j\theta}z^{-1}}\right] = \frac{A}{2}\left[\frac{z}{z - ae^{j\theta}} + \frac{z}{z - ae^{-j\theta}}\right]$$

The ROC for both terms is $|z| > |a|$ since $|e^{-j\theta}| = |e^{j\theta}| = 1$.

Example 4.11

Find the ROC of the z-transform of $x(n) = (.5)^n u(n) + (.4)^n u(n)$.

SOLUTION

From the properties of the z-transform (which will be discussed later), we have

$$X(z) = X_1(z) + X_2(z)$$

where $X_1(z)$ and $X_2(z)$ are the z-transforms for $(.5)^n u(n)$ and $(.4)^n u(n)$, respectively.

$$X(z) = \frac{1}{1 - .5z^{-1}} + \frac{1}{1 - .4z^{-1}} = \frac{z}{z - .5} + \frac{z}{z - .4} = \frac{z(z - .4) + z(z - .5)}{(z - .5)(z - .4)}$$

The ROC is given by $|z| > .5$ and $|z| > .4$. Thus, we conclude that the ROC is $|z| > .5$. This is the annular region outside the circle of radius .5 in magnitude.

Example 4.12

Find the ROC of the z-transform of $x(n) = (.5)^n u(n) + (.9)^n u(-n - 1)$.

SOLUTION

The first signal, $x(n) = (.5)^n u(n)$, has the transform

$$X_1(z) = \sum_{n-0}^{\infty} (.5)^n z^{-n} = \frac{1}{1 - .5z^{-1}} = \frac{z}{z - .5}$$

with the ROC $|z| > .5$. The second signal $x_2(n) = (.9)^n u(-n - 1)$ has the z-transform

$$X_2(z) = \sum_{n=-\infty}^{-1} (.9)^n z^{-n}$$

Let as add 1 and subtract 1 from the right side of this equation so that we make the summation stop at $n = 0$. We will get

$$X_2(z) = \sum_{n=-\infty}^{n=0} (.9)^n z^{-n} - 1$$

To start the summation from $n = 0$ to $n = \infty$, we replace n by $-n$ under the summation to get

$$X_2(z) = \sum_{n=0}^{\infty} (.9)^{-n} z^n - 1 = \sum_{n=0}^{\infty} (.9^{-1} z^1)^n - 1 = \frac{1}{1-.9^{-1}z} - 1 = -\frac{1}{1-.9z^{-1}}$$

For the second signal, we require that $|(.9)^{-1} z^1| < 1$. This implies an ROC given by $|z| < .9$. Thus, to find the ROC of $X(z)$, we require that $|z| > .5$ and $|z| < .9$. This means that the ROC is $5 < |z| < .9$.

4.5   INVERSE z-TRANSFORM

The inverse z-transform can be obtained analytically as

$$x(n) = \frac{1}{2\pi j} \oint_c X(z) z^{n-1} \, dz \qquad (4.4)$$

with $j = \sqrt{-1}$ and c is a counterclockwise closed path in the z-plane.

To avoid integration in the z-plane to find $x(n)$ from $X(z)$, we can use other ways to find $x(n)$ given $X(z)$ with the help of Tables 4.1 and 4.2.

4.5.1   PARTIAL FRACTION EXPANSION

We will assume that our signals $x(n)$ are defined for $n > 0$. The best way to illustrate the method is to give an example.

Example 4.13

Consider the signal in the z-domain

$$X(z) = \frac{z}{(z-1)(z-2)}$$

What is $x(n)$?

TABLE 4.1
Selected z-Transform Pairs

$x(n)$	$X(z)$	ROC
$A\delta(n)$	A	Entire z-plane
$Au(n)$	$\dfrac{Az}{z-1}$	$\|z\| > 1$
$nu(n)$	$\dfrac{z}{(z-1)^2}$	$\|z\| > 1$
$a^n u(n)$	$\dfrac{z}{z-a}$	$\|z\| > \|a\|$
$na^n u(n)$	$\dfrac{az}{(z-a)^2}$	$\|z\| > \|a\|$
$n^2 u(n)$	$\dfrac{z(z+1)}{(z-1)^3}$	$\|z\| > 1$
$n^2 a^n u(n)$	$\dfrac{az(z+a)}{(z-a)^3}$	$\|z\| > \|a\|$
$A\sin(\theta n)u(n)$	$\dfrac{Az\sin\theta}{z^2 - 2z\cos\theta + 1}$	$\|z\| > \|a\|$
$A\cos(\theta n)u(n)$	$\dfrac{z(z-\cos\theta)}{z^2 - 2z\cos\theta + 1}$	$\|z\| > 1$
$a^n \sin(\theta n)u(n)$	$\dfrac{az\sin\theta}{z^2 - 2az\cos\theta + a^2}$	$\|z\| > \|a\|$

TABLE 4.2
z-Transform Properties

Discrete Time Domain	z-Domain
$a_1 x_1(n) + a_2 x_2(n)$	$A_1 X_1(z) + a_2 X_2(z)$
$x(n - n_0)u(n - n_0)$	$z^{-n0}\, X(z)\ n_0 \geq 0$
$a^n\, x(n)$	$X\left(\dfrac{z}{a}\right)$
$n\, x(n)$	$-z\dfrac{dX(z)}{dz}$
$x\left(\dfrac{n}{p}\right)$	$X(z^p)$ for positive p
$x_1(n) * x_2(n)$	$X_1(z)\, X_2(z)$
$x(0)$	$\lim\limits_{z \to \infty} X(z)$
$x(\infty)$	$\lim\limits_{z \to \infty}(z-1)X(z)$ for known $x(\infty)$

SOLUTION

$X(z)$ can be written as

$$X(z) = \frac{A}{z-1} + \frac{B}{z-2}$$

where A and B are constants. After determining A and B, we will use the entries in Table 4.1 and the properties in Table 4.2 to find $x(n)$. However, the entry

$$a^n u(n) \leftrightarrow \frac{z}{z-a}$$

requires a z in the numerator. Therefore, we need to divide $X(z)$ by z then do the partial fraction expansion. After we are done, we will again multiply the results by z to get $X(z)$. So we will write

$$\frac{X(z)}{z} = \frac{z}{z(z-1)(z-a)} = \frac{1}{(z-1)(z-2)} = \frac{A}{z-1} + \frac{B}{z-2}$$

The constants are determined as

$$A = \frac{1}{z-2}\bigg|_{z=1} = -1$$

$$B = \frac{1}{z-1}\bigg|_{z=2} = 1$$

Therefore,

$$\frac{X(z)}{z} = \frac{-1}{z-1} + \frac{1}{z-2}$$

The z-transform is then

$$X(z) = \frac{-z}{z-1} + \frac{z}{z-2}$$

Now we can use Table 4.1 to get $x(n)$ with the help of entry 1 in Table 4.2. We will get

$$x(n) = -(1)^n u(n) + (2)^n u(n) = -u(n) + (2)^n u(n)$$

with

$$x(0) = -1 + 1 = 0$$

$$x(1) = -1 + 2 = 1$$

$$x(2) = -1 + 4 = 3$$

$$x(3) = -1 + 8 = 7$$

4.5.2 LONG DIVISION

In most cases, $X(z)$ can be put in a rational fraction form as a ratio of two polynomials in z, the numerator and the denominator. Then, we can use long division to find the first few values of $x(n)$. This method is good to check the results of the closed form for $x(n)$. The two polynomials should be put in descending power of z.

Example 4.14

Find the first three values of $x(n)$ for

$$X(z) = \frac{z}{(z-1)(z-2)}$$

SOLUTION

We can multiply out the denominator to get

$$X(z) = \frac{z}{z^2 - 3z + 2}$$

Now we arrange $X(z)$ as in the following:

$$
\begin{array}{r}
z^{-1} + 3z^{-2} + 7z^{-3} + \cdots \\
z^2 - 3z + 2 \overline{)\ z} \\
+ \\
-z + 3 - 2z^{-1} = 3 - 2z^{-1} \\
+ \\
-3 + 9z^{-1} - 6z^{-2} = 7z^{-1} - 6z^{-2} \\
+ \\
-7z^{-1} + 21z^{-2} - 14z^{-3} = 15z^{-2} - 14z^{-3} \\
\vdots
\end{array}
$$

The result is read as

$$X(z) = \frac{z}{z^2 - 3z + 7} = 0z^0 + (1)z^{-1} + 3z^{-2} + 7z^{-3} + \cdots$$

This indicates that

$$x(0) = 0, \quad x(1) = 1, \quad x(2) = 3, \quad \text{and} \quad x(3) = 7$$

as was obtained before with the closed form.

4.6 PROPERTIES OF THE z-TRANSFORM

Next, we will discuss some of the important properties of the z-transform. We assume that the signals start at $n \geq 0$.

4.6.1 LINEARITY PROPERTY

The z-transform of $x(n) = a_1 x_1(n) + a_2 x_2(n)$ is

$$x(z) = \sum_{n=0}^{\infty} \left(a_1 x_1(n) + a_2 x_2(n) \right) z^{-n}$$

$$= a_1 \sum_{n=0}^{\infty} x_1(n) z^{-n} + a_2 \sum_{n=0}^{\infty} x_2(n) z^{-n}$$

$$= a_1 x_1(z) + a_2 X_2(z)$$

Therefore, we write

$$x(n) = a_1 x_1(n) + a_2 x_2(n) \leftrightarrow a_1 X_1(z) + a_2 X_2(z)$$

4.6.2 SHIFTING PROPERTY

The z-transform of $x(n - n_0)\, u(n - n_0)$ is

$$\sum_{n=0}^{\infty} x(n - n_0) u(n - n_0) z^{-n}$$

Since $u(n - n_0) = 1$ for $n \geq n_0$. We have the z-transform of $x(n - n_0)\, u(n - n_0)$ written as

$$\sum_{n=n_0}^{\infty} x(n - n_0) z^{-n}$$

Let $m = n - n_0$, then

$$\sum_{n=n_0}^{\infty} x(n - n_0) z^{-n} = \sum_{m=0}^{\infty} x(m) z^{-(m+n_0)} = z^{-n_0} \sum_{m=0}^{\infty} x(m) z^{-n_0} X(z)$$

So we write

$$x(n - n_0) u(n - n_0) \leftrightarrow z^{-n_0} X(z)$$

The z-transform of $x(n - n_0)$ is

$$\sum_{n=0}^{\infty} x(n - n_0)z^{-n}$$

Let $n - n_0 = m$. Then the transform of $x(n - n_0)$ is

$$\sum_{m=-n_0}^{\infty} x(m)z^{-(m+n_0)} = \sum_{m=-n_0}^{\infty} x(m)z^{-m}z^{-n_0} = \left[\sum_{m=-n_0}^{-1} x(m)z^{-m}z^{-n_0} + \sum_{m=0}^{\infty} x(m)z^{-m}z^{-n_0} \right]$$

$$\sum_{m=-n_0}^{\infty} x(m)z^{-(m+n_0)} = \left[\sum_{m=-n_0}^{-1} x(m)z^{-m}z^{-n_0} + X(z)z^{-n_0} \right]$$

Therefore, we write

$$x(n - n_0) \leftrightarrow \left[\sum_{m=-n_0}^{-1} x(m)z^{-m}z^{-n_0} + X(z)z^{-n_0} \right]$$

The reason for this derivation is to take into account the initial conditions for $y(n)$ since they are always given for $n < 0$.

The z-transform of $x(n + n_0)$ is

$$\sum_{n=0}^{\infty} x(n + n_0)z^{-n}$$

Let $n + n_0 = m$. Then the transform of $x(n + n_0)$ is

$$\sum_{m=n_0}^{\infty} x(m)z^{-(m-n_0)} = \sum_{m=n_0}^{\infty} x(m)z^{-m}z^{n_0} = \left[\sum_{m=0}^{\infty} x(m)z^{-m}z^{n_0} - \sum_{m=0}^{m=n_0-1} x(m)z^{-m}z^{n_0} \right]$$

$$= z^{n_0} X(z) - \sum_{m=0}^{m=n_0-1} x(m)z^{-m}z^{n_0}$$

Therefore, we write

$$x(n + n_0) \leftrightarrow z^{n_0} X(z) - \sum_{m=0}^{m=n_0-1} x(m)z^{-m}z^{n_0}$$

The reason for this derivation will be apparent when we talk about state-space systems in later chapters. This also takes into consideration nonzero initial conditions.

4.6.3 MULTIPLICATION BY e^{-an}

The z-transform of $e^{-an} u(n)x(n)$ is

$$\sum_{n=0}^{\infty} e^{-an} x(n) z^{-n} = \sum_{n=0}^{\infty} x(n) \left(e^{+a} z^{+1} \right)^{-n}$$

Thus, we have

$$e^{-an} x(n) \leftrightarrow X\left(e^{a} z \right)$$

4.6.4 CONVOLUTION

The z-transform of

$$x_1(n) * x_2(n) = \sum_{k=-\infty}^{+\infty} x_1(k) x_2(n-k)$$

if $x_1(n)$ and $x_2(n)$ start at $n \geq 0$ is obtained using the z-transform defining summation equation

$$\sum_{n=0}^{\infty} \left[\sum_{k=0}^{\infty} x_1(k) x_2(n-k) \right] z^{-n}$$

which is also given as

$$\sum_{k=0}^{\infty} x_1(k) \sum_{n=0}^{\infty} x_2(n-k) z^{-n}$$

Let $n - k = m$ in the inner summation. Then $m = -k$ will be the lower summation. But $x_2(m)$ is defined only for $m \geq 0$. Therefore,

$$\sum_{k=0}^{\infty} x_1(k) \sum_{m=0}^{\infty} x_2(m) z^{-m-k} = \sum_{k=0}^{\infty} x_1(k) z^{-k} \sum_{m=0}^{\infty} x_2(m) z^{-m} = X_1(z) X_2(z)$$

From this, we write

$$x_1(n) * x_2(n) \leftrightarrow X_1(z) X_2(z)$$

This is the convolution equation in real time related to the convolution in the frequency domain. We see that convolution, a sometimes difficult operation, in real time is simply complex multiplication in the z-domain.

4.7 REPRESENTATION OF TRANSFER FUNCTIONS AS BLOCK DIAGRAMS

Consider the general third-order transfer function:

$$\frac{Y(z)}{X(z)} = \frac{az^3 + bz^2 + cz + d}{z^3 + ez^2 + fz + g}$$

The block diagram representation for this transfer function is obtained using the following steps.

1. The system is third order and hence we need three delay elements. Each delay will be preceded by a summer and followed by a summer. The initial diagram is shown in Figure 4.2, where z^{-1} represents a delay element.
2. Next we feed forward, a to the summer at the output $Y(z)$, b to the summer before the third delay, c to the summer before the second delay, and d to the summer before the first delay. The modified picture is shown in Figure 4.3.
3. Now we feed backward, $-g$ to the summer before the first delay, $-f$ to the summer before the second delay, and $-e$ to the summer before the third delay. The final block is shown in Figure 4.4.

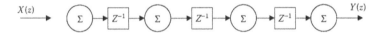

FIGURE 4.2   Block diagram step 1.

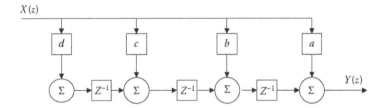

FIGURE 4.3   Block diagram step 2.

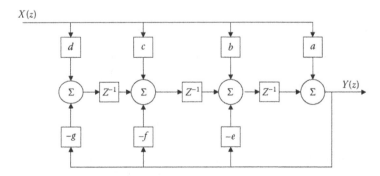

FIGURE 4.4   Block diagram step 3.

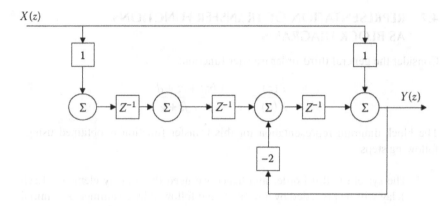

FIGURE 4.5 Block diagram for Example 4.15.

Example 4.15

Draw the block diagram for the system described by the following transfer function:

$$\frac{Y(z)}{X(z)} = \frac{z^3 + 1}{z^3 + 2z^2}$$

SOLUTION

First we write the system transfer function as

$$\frac{Y(z)}{X(z)} = \frac{z^3 + 0z^2 + 0z + 1}{z^3 + 2z^2 + 0z + 0}$$

We see that we need three delay elements. The block is shown in Figure 4.5. Note that we have only one feedback line to the summer that precedes the third delay. Also we have only two forward paths as seen in the figure.

4.8 $x(n)$, $h(n)$, $y(n)$, AND THE z-TRANSFORM

Throughout this book, we have been using $x(n)$ to represent the input, $h(n)$ to represent the impulse response, and $y(n)$ to represent the output. In discrete real time, the output is given using the convolution.

$$y(n) = x(n) * h(n)$$

But as we saw earlier, convolution is multiplication is the transform domain. Thus, we write $Y(z) = X(z) H(z)$. We then can use Tables 4.1 and 4.2 to get back to the real-time signal $y(n)$.

Example 4.16

If $x(n) = u(n)$ is an input to the system with $h(n) = (.5)^n u(n)$, what is the output $y(n)$?

SOLUTION

The output in the z-domain is

$$Y(z) = X(z)H(z)$$

The input in the z-domain is $X(z) = z/(z - 1)$ and the impulse response is $H(z) = z/(z - .5)$. Thus, using the convolution property of the z-transform, we write

$$Y(z) = \frac{z}{z-1}\frac{z}{z-.5}$$

We will find $Y(z)/z$ first to make use of Table 4.1.

$$\frac{Y(z)}{z} = \frac{z}{(z-1)(z-5)} = \frac{A}{z-1} + \frac{B}{z-.5}$$

The constants A and B are calculated as

$$A = \frac{z}{z-\frac{1}{2}}\Bigg|_{z=1} = \frac{1}{1-\frac{1}{2}} = 2$$

$$B = \frac{z}{z-1}\Bigg|_{z=\frac{1}{2}} = \frac{1/2}{1/2-1} = -1$$

Therefore,

$$\frac{Y(z)}{z} = \frac{2}{z-} - \frac{1}{z-.5}$$

and

$$Y(z) = \frac{2z}{z-1} - \frac{z}{z-.5}$$

To get back $y(n)$, we use Table 4.1 and write

$$y(n) = 2u(n) - .5^n u(n)$$

4.9   SOLVING DIFFERENCE EQUATION USING THE z-TRANSFORM

We have seen before that the z-transform of $x(n - n_0)$ is

$$x(n-n_0) \leftrightarrow \left[\sum_{m=-n_0}^{-1} x(m)z^{-m}z^{-n_0} + X(z)z^{-n_0} \right]$$

This relation is important when we z-transform difference equation.

Example 4.17

Consider the system

$$y(n) = .5y(n-1) = x(n)$$

with $y(-1) = 0$ and $x(n) = u(n)$. Find $y(n)$ for $n \geq 0$.

SOLUTION

We will z-transform the given difference equation term by term.

$$y(n) \leftrightarrow Y(z)$$

$$y(n-1) \leftrightarrow \sum_{m=-1}^{-1} y(m)z^{-1}z^{-m} + z^{-1}Y(z) = y(-1)z^{-1}z^{+1} + z^{-1}Y(z)$$

Therefore, the given equation becomes

$$Y(z) - .5\left[y(-1)z^0 + z^{-1}Y(z) \right] = X(z) = \frac{z}{z-1}$$

or

$$Y(z) - .5z^{-1}Y(z) = \frac{z}{z-1}$$

Solving for Y(z), we get

$$Y(z) = \frac{z^2}{(z-1)(z-.5)}$$

By doing partial fraction expansion on Y(z), we get

$$Y(z) = \frac{2z}{z-1} - \frac{z}{z-.5}$$

and

$$y(n) = 2u(n) - (.5)^n u(n) \quad n \geq 0$$

Example 4.18

Use the z-transform to find the impulse response of the system

$$y(n) - y(n-2) = x(n)$$

SOLUTION

We z-transform the previous equation term by term to get

$$y(z) - \left[y(-2)z^0 + y(-1)z^{-1} + z^{-2}y(z) \right] = X(z)$$

With

$$X(z) = 1 \leftrightarrow \delta(n) = x(n)$$

and $y(-2) = y(-1) = 0$ (calculating the impulse response), we have

$$Y(z)\left(1 - z^{-2}\right) = X(z)$$

or

$$Y(z) = \frac{1}{1 - z^{-2}} = \frac{z^2}{z^2 - 1}$$

and

$$\frac{Y(z)}{z} = \frac{z}{z^2 - 1} = \frac{A}{z-1} + \frac{B}{z+1}$$

The constants are

$$A = \frac{z}{z+1}\bigg|_{z=1} = \frac{1}{2}$$

$$B = \frac{z}{z+1}\bigg|_{z=-1} = \frac{1}{2}$$

Thus,

$$Y(z) = \frac{1/2z}{z-1} + \frac{1/2z}{z+1}$$

and the output $y(n) = h(n)$ is

$$y(n) = \frac{1}{2}\left[u(n) + (-1)^n u(n) \right]$$

4.10   CONVERGENCE REVISITED

Consider the following transfer function in the z-domain:

$$H(z) = \frac{z^2 + z + 1}{z^2 + 2z + 1}$$

Let us first find $h(n)$ using long division by first putting the numerator and the denominator of the transfer function in the descending powers of z. In this case, we will get

$$H(z) = 1z^0 - z^{-1} + 2z^{-2} - 3z^{-3} + 4z^{-4} + \cdots$$

and the impulse response $h(n)$ in this case is

$$h(n) = \delta(n) - \delta(n-1) + 2\delta(n-2) - 3\delta(n-3) + 4\delta(n-4) + \cdots$$

This impulse response is causal since $h(n)$ has zero values for $n < 0$. However, if we put $H(z)$ in the form

$$H(z) = \frac{1 + z + z^2}{1 + 2z + z^2}$$

we will get

$$H(z) = 1 - z + 2z^2 - 3z^3 + 4z^4 + \cdots$$

and $h(n)$ in this case is

$$h(n) = \delta(n) - \delta(n+1) + 2\delta(n+2) - 3\delta(n+3) + 4\delta(n+4) + \cdots$$

The signal is noncausal because it is zero for $n > 0$ and nonzero or $n \leq 0$.
   Notice that

$$H(z) = \frac{1 + z + z^2}{1 + 2z + z^2} = \frac{z^2 + z + 1}{z^2 + 2z + 1}$$

But we have two signals as follows:

$$h(n) = \delta(n) - \delta(n+1) + 2\delta(n+2) - 3\delta(n+3) + 4\delta(n+4) + \cdots$$
$$h(n) = \delta(n) - \delta(n-1) + 2\delta(n-2) - 3\delta(n-3) + 4\delta(n-4) + \cdots$$

There is a reason for what we see here. Consider the signal $h(n) = a^n u(n)$ with its z-transform:

$$H(z) = \sum_{n=-\infty}^{+\infty} a^n u(n) z^{-n} = \sum_{z=0}^{\infty} a^n z^{-n} = \frac{1}{1 - az^{-1}} = \frac{z}{z - a}$$

This $H(z)$ has the ROC $|z| > |a|$ since we require that $|az^{-1}| < 1$ for the series to be summable. Notice also that $H(z)$ has its pole at $z = a$ and that $H(z)$ converges outside the circle of radius of magnitude a.

Consider next the signal $h(n) = a^n u(-n - 1)$ with its z-transform given by

$$H(z) = \sum_{-\infty}^{+\infty} a^n u(-n-1) z^{-n} = \sum_{n=-\infty}^{-1} a^n z^{-n} = \sum_{n=-\infty}^{n=0} a^n z^{-n} - 1 = \sum_{n=0}^{n=\infty} a^{-n} z^n - 1$$

where we changed the sign on n when we changed the limit to start at $n = 0$ and end at $n = \infty$. $H(z)$ then is written as

$$H(z) = \frac{1}{1-a^{-1}z} - 1 = \frac{1-\left(1-a^{-1}z\right)}{1-a^{-1}z} = \frac{a^{-1}z}{1-a^{-1}z} = \frac{z}{z-a}$$

But the ROC is now different since in this case we required that $|a^{-1}z| < 1$ for the series to converge. This means that $|z/a| < 1$ or $|z| < |a|$.

From what has been discussed it is clear that to find $h(n)$ from $H(z)$, the ROC must be given. To generalize, let us consider that $H(z)$ has N poles and M zeros. If $H(z)$ has an ROC for which

$$|z| > |p_i|$$

When p_i is the ith pole that is farthest from the pole at the origin, then in this case the system is causal. But if the ROC is

$$|z| < |p_i|$$

where p_i is the ith pole that is closest to the pole at the origin, then in this case the system is noncausal. Let us look at some examples.

Example 4.19

Consider the system

$$H(z) = \frac{(2z-3)z}{(z-1)(z-2)}$$

with ROC $|z| > 2$. Find $h(n)$.

SOLUTION

The transfer function can be written as

$$H(z) = \frac{(2z-3)z}{(z-1)(z-2)} = \frac{z}{z-1} + \frac{z}{z-2}$$

We can see that the ROC of $H(z)$ is outside the rings $|z| = 1$ and $|z| = 2$ and hence $h(n) = (1)^n u(n) + (+2)^n u(n)$.

Example 4.20

Consider the same $H(z)$ as in Example 4.19, but the ROC now is $|z| < 1$. What is $h(n)$?

SOLUTION

We can see in this case that the ROC is inside both rings $|z| = 1$ and $|z| = 2$. Therefore,

$$h(n) = -(1)^n u(-n-1) + -(2)^n u(-n-1)$$

The minus sign here is added because the system is noncausal.

4.11   FINAL-VALUE THEOREM

The final value of the signal $x(n)$ as $n \to \infty$ can be obtained using the z-transform. The derivation is omitted since it is somewhat involved. We have

$$x(\infty) = \lim_{n \to \infty} x(n) = \lim_{z \to 1} (z-1) X(z) \tag{4.5}$$

if $x(n)$ has a final value. $x(n)$ will have a final value if all the poles of $X(z)$ are within the unit circle. This is to say that for all the poles z_i, $|z_i| < 1$.

Example 4.21

Consider the system

$$h(n) = (.5)^n u(n)$$

Find $h(\infty)$ or the final value of $h(n)$.

SOLUTION

We can see here, since the expression for $h(n)$ is simple, that $h(\infty) = (.5)^\infty u(\infty) = 0$. This is clear because $(.5)^\infty$ will approach zero. But we can use the final-value theorem since the pole for $H(z) = z/(z - .5)$ is within the unit circle. Therefore,

$$h(\infty) = \lim_{z \to 1} (z-1) \left(\frac{z}{z-.5} \right) = (1-1) \frac{1}{1-.5} = 0$$

4.12   INITIAL-VALUE THEOREM

The initial-value theorem is used to find the initial value $x(0)$ for the signal $x(n)$. From the z-transform of $x(n)$, we write

$$X(z) = x(0) + x(1)z^{-1} + z(2)z^{-2} + \cdots$$

If we take the limit of $X(z)$ as z approaches infinity, we will have

$$\lim_{z \to \infty} X(z) = x(0) + \lim_{z \to \infty} \frac{x(1)}{z} + \lim_{z \to \infty} \frac{x(2)}{z^2} + \cdots$$

So we have

$$\lim_{z \to \infty} X(z) = x(0) \qquad (4.6)$$

as the initial-value theorem

Example 4.22

If a certain system has the impulse response

$$h(n) = (.5)^n u(n)$$

what would be $h(0)$, the initial value for $h(n)$?

SOLUTION

It is clear that $h(0) = (.5)^0 \, u(0) = 1$. Using the initial-value theorem, we have

$$h(0) = \lim_{z \to \infty} H(z) = \lim_{z \to \infty} \frac{z}{z - .5} = 1$$

4.13   SOME INSIGHTS: POLES AND ZEROES

The transfer function $H(z)$ of a linear time-invariant system is a very important representation. It tells us many things about the stability of the system, the poles, the zeros, and the shape of the transients of the output of the system. Using $H(z)$ we can find the steady-state response of the system and the particular solution of the system all in one shot.

4.13.1   Poles of the System

The poles of the system are the roots of the denominator, the algebraic equation in the variable z of the transfer function $H(z)$

$$H(z) = \frac{Nl(z)}{D(z)}$$

$D(z)$ is a polynomial in z of order equal to the order of the system. The roots of the denominator $D(z)$ are called the poles of the system. These are the same poles we discussed in Chapter 2. We called them then the eigenvalues of the system $D(z)$ is actually the characteristic equation of the system or, as referred to before, the auxiliary equation of the system.

14.13.2   ZEROS OF THE SYSTEM

The roots of the numerator $N(z)$ are called the zeros of the system.

4.13.3   STABILITY OF THE SYSTEM

The poles of the system determine its stability. If the poles are *all* within the unit circle, then the system at hand is stable and the transients will die as time progresses. The stability of the system is determined by the poles and not the zeros. If one of the poles is outside the unit circle, then the system is not stable. You may have zeros that are outside the unit circle, but the location of the zeros has no effect on the stability of the system.

Given $H(z)$, the roots of the denominator will determine the general shape of the output $y(n)$, which, in this case, is $h(n)$ because the input $x(n)$ is the impulse $\delta(n)$. If $D(z)$ has two roots (second-order system) called α_1 and α_2, then the output will have the general form

$$y(n) = h(n) = c_1(\alpha_1)^n + c_2(\alpha_2)^n$$

where the constant c's are to be determined. The exponential terms will determine the shape of the transients. If one of the α's is outside the unit circle, the output will grow without bounds. If the two α's are within the unit circle, the output will die as time progresses. The α's are the eigenvalues or the poles of the system.

The transfer function $Y(z)/X(z)$ is called $H(z)$ if the input $X(z)$ is 1 ($x(n) = \delta(n)$). The transfer function $Y(z)/X(z)$ is very important as we will see later in the design of linear time-invariant systems.

4.14   END-OF-CHAPTER EXAMPLES

EOCE 4.1

Find the z-transform of the signals

1. $x(n) = \left(\dfrac{1}{3}\right)^n u(n)$

2. $x(n) = \left(\dfrac{1}{2}\right)^n u(-n-1)$

3. $x(n) = \left(\dfrac{1}{2}\right)^n u(n) - \left(\dfrac{1}{2}\right)^n u(-n-1)$

and indicate their ROC.

SOLUTION

For the first signal $x(n) = (1/3)^n\, u(n)$, the z-transform is

$$X(z) = \sum_{n=-\infty}^{+\infty} \left(\frac{1}{3}\right)^n u(n) z^{-n} = \sum_{n=0}^{\infty} \left(\frac{1}{3} z^{-1}\right)^n = \frac{1}{1-(1/3)z^{-1}} = \frac{z}{z-(1/3)} \text{ for } \left|\frac{1}{3} z^{-1}\right| < 1$$

or $|1/3| < |z|$. So the ROC in $|z| > 1/3$.

For the second signal, $x(n) = -(1/2)^n u(-n-1)$, the z-transform is given by

$$X(z) = \sum_{n=-\infty}^{+\infty} \left(\frac{1}{2}\right)^n u(-n-1) z^{-n} = -\sum_{n=-\infty}^{-1} \left(\frac{1}{2}\right)^n z^{-n}$$

$$X(z) = -\sum_{n=1}^{n=\infty} \left(\frac{1}{2}\right)^{-n} z^n = 1 - \sum_{n=0}^{\infty} \left(\left(\frac{1}{2}\right)^{-1} z\right)^n = 1 - \frac{1}{1-(1/2)^{-1}z} = \frac{z}{z-(1/2)}$$

with the ROC $|z| < 1/2$.

For the last signal, $x(n) = (1/3)^n\, u(n) - (1/2)^n\, u(-n-1)$, the z-transform is

$$X(z) = \sum_{n=-\infty}^{+\infty} \left(\frac{1}{3}\right)^n u(n) z^{-n} - \sum_{n=-\infty}^{+\infty} \left(\frac{1}{2}\right)^n u(-n-1) = \frac{z}{z-(1/3)} + \frac{z}{z-(1/2)}$$

with ROC now as $|z| > 1/3$ and $|z| < 1/2$. By combining these conditions, we get

$$\frac{1}{3} < |z| < \frac{1}{2}$$

for the ROC.

EOCE 4.2

What is the z-transform of the signal

$$x(n) = (n+2)(.5)^n u(n)$$

SOLUTION

$x(n)$ can be written as

$$x(n) = nx_1(n) + 2x_1(n)$$

where $x_1(n) = (.5)^n u(n)$. Thus, we have

$$X_1(z) = \frac{z}{z-.5} \text{ with } |z| > .5$$

Next, we can use the differentiation property in Table 4.2 to find $X(z)$ as

$$X(z) = -z\frac{d}{dz}X_1(z) + 2\frac{z}{z-.5} = -z\left[\frac{(z-.5)-z}{(z-.5)^2}\right] + 2\frac{z}{z-.5}$$

$$X(z) = \frac{.5z}{(z-.5)^2} + 2\frac{z}{z-.5} = \frac{.5z + 2z(z-.5)}{(z-.5)^2} = \frac{2z^2 - .5z}{(z-.5)^2}$$

with $|z| > .5$ as the ROC.

EOCE 4.3

What is $X(z)$ if

$$x(n) = \cos(n)u(n) + nu(n)$$

SOLUTION

We can divide the given signal into two parts.

$$X(z) = X_1(z) + X_2(z)$$

where $X_1(z)$ and $X_2(z)$ are the z-transforms of $\cos(n)u(n)$ and $nu(n)$, respectively. Therefore,

$$X(z) = \frac{1 - \cos(1)z^{-1}}{1 - 2\cos(1)z^{-1} + z^{-2}} + \frac{z^{-1}}{(1 - z^{-1})^2}$$

with $|z| > 1$ as the ROC. $X_2(z)$ can be obtained by finding the z-transform of $u(n)$ first, then taking the derivative with respect to z as

$$X_2(z) = \frac{-zd}{dz}\left(\frac{z}{z-1}\right) = -z\left(\frac{z-1-z}{(z-1)^2}\right) = \frac{z}{(z-1)^2} = \frac{z^{-1}}{(1 - z^{-1})^2}$$

EOCE 4.4

What is the z-transform of the signal

$$x(n) = x_1(n) * x_2(n)$$

with

$$x_1(n) = \delta(n) + 2\delta(n-1)$$

and

$$x_2(n) = \delta(n-1) + 3\delta(n-2)$$

SOLUTION

By using the convolution property, we write

$$X(z) = X_1(z)X_2(z) = (1 + 2z^{-1})(z^{-1} + 3z^{-2}) = z^{-1} + 3z^{-2} + 2z^{-2} + 6z^{-3}$$

and x(n) then is

$$x(n) = \delta(n-1) + 5\delta(n-2) + 6\delta(n-3)$$

We can also use MATLAB® and convolve $x_1(n)$ with $x_2(n)$ and write the script EOCE4_4
to get

$$x = \begin{matrix} \{0 & 1 & 2 & 3\} \\ \uparrow & & & \end{matrix}$$

which indicates that

$$X(z) = 0z^0 + 1z^{-1} + 5z^{-2} + 6z^{-3}$$

EOCE 4.5

With

$$x_1(n) = \delta(n+1) + \delta(n) + \delta(n-1)$$

and

$$x_2(n) = \delta(n) + \delta(n-1)$$

what is the z-transform if $x(n) = x_1(n) * x_2(n)$?

SOLUTION

Convolution in real time is multiplication in the z-domain. Thus,

$$X(z) = X_1(z)X_2(z) = (z + 1 + z^{-1})(1 + z^{-1})$$
$$= z + 1 + 1 + z^{-1} + z^{-1} + z^{-2} = z + 2 + 2z^{-1} + z^{-2}$$

and the inverse z-transform is

$$x(n) = \delta(n+1) + 2\delta(n) + 2\delta(n-1) + \delta(n-2)$$

We can also use MATLAB to do this but with some attention to the starting index of each signal. We use the MATLAB function conv if the two signals start at $n = 0$. In our case, $x_1(n)$ starts at $n = -1$ and $x_2(n)$ starts at $n = 0$. In this case, we have to fix the starting and ending indices to find $x(n)$. We write the MATLAB script EOCE4_5 to do that.

We will get

$$
\begin{array}{cc}
-1 & 1 \\
0 & 2 \\
1 & 2 \\
2 & 1
\end{array}
$$

and from this, we have

$$x(n) = \delta(n+1) + 2\delta(n) + 2\delta(n-1) + \delta(n-2)$$

EOCE 4.6

When

$$x(n) = (.5)^n u(n) + (.3)^n u(n) + (.9)^n u(n)$$

what is $X(z)$ and its ROC?

SOLUTION

Using the linearity property of the z-transform, we write

$$
\begin{aligned}
X(z) &= \frac{z}{z-.5} + \frac{z}{z-.3} + \frac{z}{z-.9} \\
&= \frac{z(z-.3)(z-.9) + z(z-.5)(z-.9) + z(z-.5)(z-.3)}{(z-.5)(z-.3)(z-.9)}
\end{aligned}
$$

and the ROC is $|z| > .3$ and $|z| > .5$ and $|z| > .9$ all satisfied simultaneously. The ROC becomes $|z| > .9$.

EOCE 4.7

Consider the transforms

$$X_1(z) = 1 + z^{-1} + 3z^{-2}$$

and

$$X_2(z) = 1 + 3z^{-2}$$

What is $x(n) = x_1(n) * x_2(n)$?

SOLUTION

$$x_1(n) = \delta(n) + \delta(n-1) + 3\delta(n-2)$$

and

$$x_2(n) = \delta(n) + 3\delta(n-2)$$

We can use MATLAB to find $x(n) = x_1(n) * x_2(n)$ by writing the script EOCE4_7. to get

$$x = \begin{Bmatrix} 1 & 1 & 6 & 3 & 9 \\ \uparrow & & & & \end{Bmatrix}$$

and

$$x(n) = \delta(n) + \delta(n-1) + 6\delta(n-2) + 3\delta(n-3) + 9\delta(n-4)$$

We can also use the z-transform to arrive at this result.

$$X(z) = X_1(z)X_2(z) = (1 + z^{-1} + 3z^{-2})(1 + 3z^{-2})$$
$$= 1 + 3z^{-2} + z^{-1} + 3z^{-3} + 3z^{-3} + 9z^{-4}$$

and then the inverse z-transform is

$$x(n) = \delta(n) + \delta(n-1) + 6\delta(n-2) + 3\delta(n-3) + 9\delta(n-4)$$

EOCE 4.8

Use MATLAB, long division and partial function expansion to find $h(n)$ if

$$H(z) = \frac{1}{z^2 - 3z + 2} \qquad \text{with ROC } |z| > 2$$

SOLUTION

Using MATLAB, first we need to put $X(z)$ in ascending powers of z^{-1} and write

$$H(z) = \frac{z^{-2}}{1 - 3z^{-1} + 2z^{-2}}$$

Then we will use the MATLAB function residuez that has the form

$$[r\ p\ k] = \text{residuez (num, den)}$$

where num and den are the coefficients of the numerator and the denominator of the rational z-transformed function. r is a vector that contains the residues, p is the vector that contains the poles, and k is the constant term that is nonzero if the degree of num is larger or equal to the degree of den, the numerator and the denominator of the rational z-transformed function. With this, we write the script EOCE4_81.

to get

$$r = 0.5000 \text{ and } -1.0000$$

$$p = 2 \text{ and } 1$$

$$k = 0.5$$

This gives the transfer function

$$H(z) = \frac{1}{2} - \frac{z}{z-1} + \frac{(1/2)z}{z-2}$$

from which we get

$$h(n) = \frac{1}{2}\delta(n) - u(n) + \frac{1}{2}(2)^n u(n)$$

using Table 4.1. We can see that the first few terms of h(n) are

$$h(0) = \frac{1}{2} - 1 + \frac{1}{2} = 0$$

$$h(1) = 0 - 1 + 1 = 0$$

$$h(2) = 0 - 1 + 2 = 1$$

We can also use MATLAB to find some terms of h(n). To do that, we can use the MATLAB function filter with an impulsive input. We have

$$Y(z) = X(z)H(z)$$

If $X(z) = 1$, $(h(n) = \delta(n))$, then $Y(z) = H(z)$ and $y(n)$ is $h(n)$. To do that, we write the script EOCE4_82.
to get the same result we found earlier.
Using long division, we can divide the numerator by the denominator to get

$$H(z) = \frac{1}{z^2 - 3z + 2} = 0z^0 + 0z^{-1} + z^{-2} + 3z^{-3} + \cdots$$

which clearly indicates that h(0) = 0, h(1) = 0 and h(2) = 1 as we saw earlier.

Using partial fraction expansion, we write the transfer function as

$$\frac{H(z)}{z} = \frac{1}{z(z-1)(z-2)} = \frac{A}{z} + \frac{B}{z-1} + \frac{C}{z-2}$$

with the constants evaluated as

$$A = \frac{1}{(z-1)(z-2)}\bigg|_{z=0} = \frac{1}{2}$$

$$B = \frac{1}{z(z-2)}\bigg|_{z=1} = -1$$

$$C = \frac{1}{z(z-1)}\bigg|_{z=2} = \frac{1}{2}$$

So the transfer function becomes

$$H(z) = \frac{(1/2)z}{z} + \frac{-z}{z-1} + \frac{(1/2)z}{z-2}$$

and

$$h(n) = \frac{1}{2}\delta(n) - u(n) + \frac{1}{2}(2)^2 u(n)$$

EOCE 4.9

Find $h(n)$ if $H(z)$ is

$$H(z) = \frac{1}{z^2 - .9z + .7} = \frac{z^{-2}}{1 - .9z^{-1} + .7z^{-2}}$$

Assume the resulting signal $h(n)$ is valid only for $n \geq 0$ and it is real.

SOLUTION

Since this $H(z)$ is not similar to any form in Table 4.1, we need to use partial fraction expansion on $H(z)$. We do that using MATLAB and write the script EOCE4_9 to get

$$r = -0.7143 - 0.4557i \text{ and } -0.7143 + 0.4557i$$

$$p = 0.4500 + 0.7053i \text{ and } 0.4500 - 0.7053i$$

$$k = 1.4286$$

$$\text{mag\_}r = 0.8473 \text{ and } 0.8473$$

$$\text{phase\_r} = -2.5737 \text{ and } 2.5737$$

$$\text{mag\_p} = 0.8367 \text{ and } 0.8367$$

$$\text{phase\_p} = 1.0029 \text{ and } -1.0029$$

Hence, we can see that

$$H(z) = 1.428 + \frac{0.847e^{-j2.57}}{1 - |0.8367|e^{j1.0029}z^{-1}} + \frac{0.847e^{j2.57}}{1 - |0.8367|e^{-j1.0029}z^{-1}}$$

with the ROC as $|z| > 0.8367$. Now we can use Table 4.1 to get the inverse transform $h(n)$. We write

$$h(n) = 1.428\delta(n) + 0.847e^{-j2.57}(.8367)^n e^{j(1.0029)n} + 0.847e^{j2.57}(.8367)^n e^{-j(1.0029)n}$$

These two terms are complex conjugate terms. We know that the sum of two complex conjugate terms is two times the real part of the complex number. Thus, we write

$$h(n) = 1.428\delta(n) + 2\,\text{Real}\!\left(0.847e^{-j2.257}(.8367)^n e^{j(1.0029)n}\right)$$

$$= 1.428\delta(n) + 2(0.847)(.8367)^n\,\text{Real}\!\left(e^{j(1.0029\ n-2.57)}\right)$$

After some simplifications, we get

$$h(n) = 1.428\delta(n) + \left[2(0.847)(.83657)^n \cos(1.0029n - 2.57)\right]u(n)$$

EOCE 4.10

Consider the system transfer function in the z-domain

$$H(z) = \frac{z}{z^3 - 6z^2 + 11z - 6}$$

with ROC $|z| > 3$. Find $h(n)$.

SOLUTION

First, we write $H(z)$ in the proper form so that we can use partial function to find $h(n)$. $H(z)$ then is

$$H(z) = \frac{z^{-2}}{1 - 6z^{-1} + 11z^{-2} - 6z^{-3}}$$

By putting $H(z)$ in partial fraction form, we use MATLAB and write the script EOCE4_10 to get

$$r = 0.5000, -1.0000, \text{ and } 0.5000$$

$$p = 3.0000, 2.0000, \text{ and } 1.0000$$

$$k = [\,] \text{ to indicate zero value}$$

Then the partial fraction expansion results in

$$H(z) = \frac{0.5}{1-z^{-1}} + \frac{-1}{1-2z^{-1}} + \frac{0.5}{1-3z^{-1}}$$

To bring $H(z)$ back into real time, we need to look carefully at the ROC for $H(z)$. For the first term $0.5/(1-z^{-1}) = 0.5z/(z-1)$, the pole is at $z = 1$. The ROC of $H(z)$ is outside the ring $|z| = 3$ and hence outside the ring $|z| = 1$. In this case,

$$\frac{0.5}{1-z^{-1}} \leftrightarrow 0.5u(n)$$

The second term $-1/(1-2z^{-1}) = -z/(z-2)$ has the pole $z = 2$. The ROC of $H(z)$ is outside the ring $|z| = 3$ and hence

$$\frac{-1}{1-2z^{-1}} \leftrightarrow -(2)^n u(n)$$

The last term $0.5/(1-3z^{-1}) = 0.5z/(z-3)$ has a pole at $z = 3$. The ROC of $H(z)$ is outside the $|z| = 3$ circle and $z = 3$ is within this ROC. Thus,

$$\frac{0.5}{1-3z^{-1}} \leftrightarrow 0.5(3)^n u(n)$$

Therefore, the inverse transform is

$$h(n) = 0.5u(n) - (2)^n u(n) + 0.5(3)^n u(n)$$

Note: When you use MATLAB to find $h(n)$ be careful to pay attention to the ROC of $H(z)$.

EOCE 4.11

Consider the following causal difference equation:

$$y(n) - .7y(n-1) = x(n)$$

where $y(n)$ is the output and $x(n)$ is the input.

1. Find the transfer function $H(z)$.
2. Find the transfer function $H(e^{j\theta})$.
3. Find $h(n)$, the impulse response.
4. Find $y(n)$ if $x(n) = u(n)$.

SOLUTION

1. With zero initial conditions we transform the difference equation term by term into the z-domain. We will have

$$Y(z) - .7z^{-1}Y(z) = X(z)$$

Taking $Y(z)$ as a common factor, we get

$$Y(z)\left[1 - .7z^{-1}\right] = X(z)$$

and the transfer function is

$$\frac{Y(z)}{X(z)} = \frac{1}{1 - .7z^{-1}} = \frac{z}{z - .7}$$

with ROC $|z| > .7$.
2. Since $H(z)$ converges for $|z| > .7$, the unit circle is inside the ROC and so we can find $H(e^{j\theta})$ from $H(z)$ as

$$H\left(e^{j\theta}\right) = H(z)\Big|_{z=e^{j\theta}} = \frac{e^{j\theta}}{e^{j\theta} - .7} = \frac{1}{1 - .7e^{-j\theta}}$$

3. $h(n)$ can be obtained directly from Table 4.1 as

$$h(n) = (.7)^{n} u(n)$$

4. With $h(n) = (.7)^n u(n)$ and $x(n) = u(n)$, we have $y(n) = x(n) * h(n)$. Or in z-domain, we have

$$Y(z) = \frac{z}{z-1}\frac{z}{z-.7} = \frac{z^2}{z^2 - 1.7z + .7} = \frac{1}{1 - 1.7z^{-1} + .7z^{-2}}$$

At this point, we can determine the ROC for $Y(z)$ as the intersection of region $|z| > 1$ and $|z| > .7$. Thus, the ROC for $Y(z)$ is $|z| > .7$. We can use MATLAB to find $Y(z)$ in partial fraction form by writing the script EOCE4_111 to get

$$r = 3.3333 \text{ and } -2.3333$$

$$p = 1.0000 \text{ and } 0.7000$$

Thus,

$$Y(z) = \frac{3.332}{1 - z^{-1}} + \frac{-2.3333}{1 - 0.7z^{-1}}$$

The inverse transform of this equation gives

$$y(n) = 3.332(1)^n u(n) - 2.3333(0.7)^n u(n)$$

We can use MATLAB to plot the results obtained in this example. We will use the MATLAB function freqz as

$$[H, theta] = freqz \ (num, den, N, 'whole')$$

where
H is the vector that contains all the frequency response values.
theta is the vector that contains the frequency values.
num and den are the coefficients of the numerator and the denominator of the transfer function.
whole indicates that freqz will evaluate H for the N theta values around the entire unit circle.

The MATLAB script is EOCE4_112.

The plots are shown in Figure 4.6.

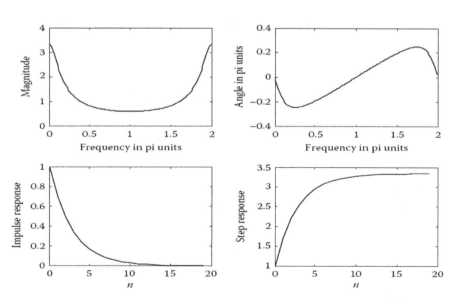

FIGURE 4.6 Signals for EOCE 4.11.

EOCE 4.12

Consider the system transfer function

$$H(z) = \frac{2z^2 - 12z}{(z - .3)(z + .2)(z - 3)}$$

Find $h(n)$ for the following ROCs and indicate stability.

1. $|z| > 3$
2. $|z| < .2$
3. $.2| z| < .3$
4. $.3 < |z| < 3$

SOLUTION

The poles of the system are at $z = .3$, $z = -.2$ and $z = 3$.

1. $|z| > 3$ is the ROC that is outside the ring $|z| = 3$. The poles at $.3$, $-.2$, and 3 are inside this ring. The system is unstable and causal with

$$h(n) = c_1 (.3)^n u(n) + c_2 (-.2)^n u(n) + c_3 (3)^n u(n)$$

2. $|z| < .2$ is the ROC that is inside the ring $|z| = .2$. The system is unstable and noncausal with

$$h(n) = c_4 (.3)^n u(-n-1) + c_5 (-.2)^n u(-n-1) + c_6 (3)^n u(-n-1)$$

3. $.2 < |z| < .3$ is the ROC that is between the two rings $|z| = .2$ and $|z| = .3$

$$h(n) = c_7 (.3)^n u(-n-1) + c_8 (-.2)^n u(n) + c_9 (3)^n u(-n-1)$$

4. $.3 < |z| < 3$ is the ROC that is between the two rings $|z| = .3$ and $|z| = 3$.

$$h(n) = c_{10} (-.2)^n u(n) + c_{11} (.3)^n u(n) + c_{12} (3)^n u(-n-1)$$

EOCE 4.13

MATLAB has a function called filter that can be used to find the response of a discrete system using initial conditions that are different from zero. The form of this function is

$$y = \text{filter}(num, den, x, y0)$$

where
   y is the output vector.
   num is the numerator coefficients vector.

den is the denominator coefficients vector.

x is the input vector.

y0 is the initial conditions vector that is derived using the actual initial conditions given to you with the system under investigation.

In real time, the Nth-order difference equation can be represented as

$$y(n) + a_1 y(n-1) + \cdots + a_N y(n-N) = b_0 x(n) + b_1 x(n-1) + \cdots + b_L x(n-L)$$

The z-transform of the this equation is

$$Y(z) = \frac{b_0 + b_1 z^{-1} + \cdots + b_L z^{-L}}{1 + a_1 z^{-1} + a_2 z^{-2} + \cdots + a_N z^{-N}}$$

The derived initial conditions for the filter function are computed as in the following.

$$y_{00} = -a_1 y(-1) - \cdots - a_N y(-N)$$
$$y_{01} = -a_2 y(-1) a_3 y(-2) - \cdots - a_N y(-N+1)$$
$$\vdots$$
$$y_{0N} = -a_N y(-1)$$

Consider the following system:

$$y(n) - y(n-1) = x(n)$$

with $y(-1) = 1$ and $x(n) = \delta(n)$. Find $y(n)$ using the z-transform.

SOLUTION

We can z-transform the difference equation term by term and write

$$Y(z) - \left[y(-1) + z^{-1} Y(z) \right] = X(z)$$

We can rearrange terms and write

$$Y(z)\left[1 - z^{-1} \right] = X(z) + y(-1)$$

We finally have

$$Y(z) = \frac{X(z)}{1 - z^{-1}} + \frac{y(-1)}{1 - z^{-1}}$$

With $X(z) = 1$, the output is

$$Y(z) = \frac{z}{z-1} + \frac{z}{z-1} + \frac{2z}{z-1}$$

Using Table 4.1, we get

$$y(n) = 2u(n)$$

Using MATLAB, we first write $Y(z)/X(z)$ without initial conditions to get

$$\frac{Y(z)}{X(z)} = \frac{1}{1-z^{-1}} = \frac{b_0}{1+(a_1)z^{-1}}$$

The initial condition vector y_0 will have one component since the system is first order. So the derived initial condition is

$$y_{00} = -a_1 y(-1) = -1(1) = -1$$

Next, we write the MATLAB script EOCE4_131 to find $y(n)$ for $x(n) = \delta(n)$.

The plots are shown in Figure 4.7.

We can also use MATLAB with the MATLAB function filtic to find the derived initial conditions for us. The general form of filtic is

$$ic = filtic\,(num,\ den,\ x0,\ y0)$$

where

    ic is the returned derived initial conditions.

    num and den are as discussed before.

    x0 and y0 are the initial condition vectors given with the problem for negative n.

We can rework the MATLAB solution using the filtic function as in the script EOCE4_132 and the result is the same.

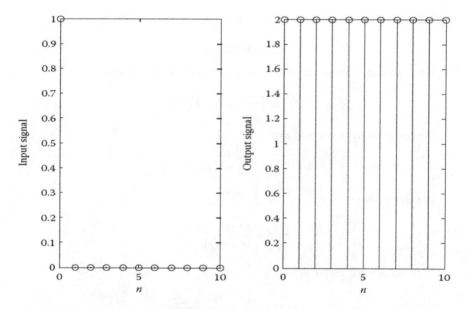

FIGURE 4.7   Signals for EOCE 4.13.

EOCE 4.14

Consider the following system:

$$y(n) - .5y(n-1) - .1y(n-2) + .2y(n-3) = x(n)$$

with $x(n) = u(n)$ and $y(-1) = 1$, $y(-2) = 2$ and $y(-3) = 3$,

1. Find the output $y(n)$ using MATLAB.
2. Find the system transfer function.
3. Is the system stable?

SOLUTION

1. The initial condition vector y_0 for the function filter is determined next.

$$y_{00} = -a_1y(-1) - a_2y(-2) - a_3y(-3) = .5(1) + .1(2) - .2(3) = .1$$
$$y_{01} = -a_2y(-1) - a_3y(-2) = .1(1) + .2(2) = .3$$
$$y_{02} = -a_3y(-1) = -.2(1) = -.2$$

We can use MATLAB to determine these derived initial conditions by writing the EOCE4_141 script. The result is

$$ic = 0.1000 - 0.3000 - 0.2000$$

With these derived initial conditions, we can write the MATLAB script EOCE4_142 to plot the response due to the step input. The plots are shown in Figure 4.8.

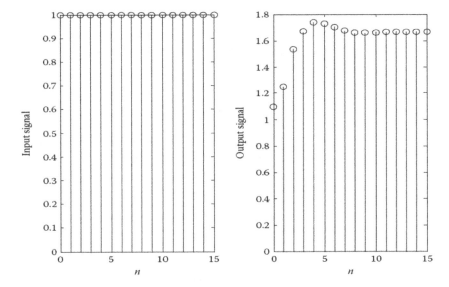

FIGURE 4.8 Signals for EOCE 4.14.

2. The system transfer function is obtained by taking the z-transform of the difference equation with zero initial conditions. The result is

$$Y(z) - .5z^{-1}zY(z) - .1z^{-2}Y(z) + .2z^{-3}Y(z) = X(z)$$

By grouping similar terms, we get

$$Y(z)\left[1 - .5z^{-1} - .1z^{-2} + .2z^{-3}\right] = X(z)$$

and the transfer function is

$$\frac{Y(z)}{X(Z)} = H(z) = \frac{1}{1 - .5z^{-1} - .1z^{-2} + .2z^{-3}}$$

3. We can use MATLAB to find the poles of H(z). We type at the MATLAB prompt

$$r = \text{roots} \left([1 - .5 - .1 .2]\right);$$

abs (r)%the magnitude of the roots
    to get 0.5000, 0.6325, and 0.6325. The roots are within the unit circle and thus the system is stable.

EOCE 4.15

Consider the system

$$y(n) = \frac{1}{2}y(n-1) = x(n) + x(n-1)$$

1. Find the impulse response h(n).
2. What is the step response?
3. Use MATLAB to find y(n) if x(n) = (.5)^n sin(n) u(n).

SOLUTION

1. If we can find H(z) for the given system, we will inverse transform H(z) to get h(n). Let us z-transform the given difference equation to get

$$Y(z) - \frac{1}{2}z^{-1}Y(z) = X(z) + z^{-1}X(z)$$

and the transfer function is then

$$\frac{Y(z)}{X(z)} = H(z) = \frac{1 + z^{-1}}{1 - (1/2)z^{-1}}$$

The ROC is $|z| > 0.5$. Using partial fraction expansion on $H(z)$, we have

$$\frac{H(z)}{z} = \frac{z+1}{z\left(z-(1/2)\right)} = \frac{A}{z} + \frac{B}{(1/2)}$$

with

$$A = \frac{z+1}{z-(1/2)}\bigg|_{z=0} = -2$$

and

$$B = \frac{z+1}{z}\bigg|_{z=1/2} = \frac{(1/2)+1}{(1/2)} = \frac{(3/2)}{(1/2)} = 3$$

Thus,

$$H(z) = \frac{-2z}{z} + \frac{3z}{z-(1/2)}$$

and from Table 4.1, we get

$$h(n) = -2\delta(n) + 3\left(\frac{1}{2}\right)^n u(n)$$

2. The step response is obtained with $X(z) = z/(z-1)$. The output will be

$$Y(z) = X(z)H(z) = \frac{z+1}{z-(1/2)}\frac{z}{z-1}$$

We will do partial fraction expansion on

$$\frac{Y(z)}{z} = \frac{z+1}{\left(z-(1/2)\right)(z-1)} = \frac{A}{z-(1/2)} + \frac{B}{z-1}$$

to get

$$A = \frac{z+1}{z-1}\bigg|_{z=\frac{1}{2}} = \frac{\left(\frac{1}{2}\right)+1}{\left(\frac{1}{2}\right)-1} = \frac{\frac{3}{2}}{-\frac{1}{2}} = -3$$

$$B = \frac{z+1}{z-(1/2)}\bigg|_{z=1} = \frac{1+1}{1-(1/2)} = \frac{2}{1/2} = 4$$

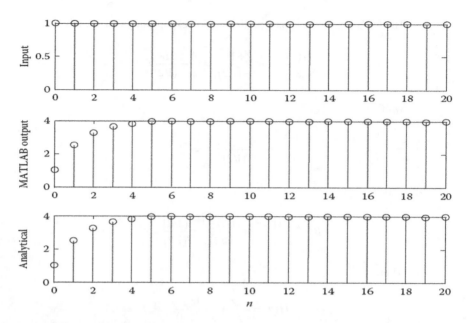

FIGURE 4.9 Signals for EOCE 4.15.

The resulting output in the z-domain is

$$Y(z) = \frac{-3z}{z-(1/2)} + \frac{4z}{z-1}$$

and the inverse z-transform is

$$y(n) = -3\left(\frac{1}{2}\right)^n u(n) + 4u(n)$$

We can use MATLAB to verify this solution by writing the script EOCE4_151.

The plots are shown in Figure 4.9.

3. We will use the MATLAB script EOCE4_152 to do that.

The plots are shown in Figure 4.10.

EOCE 4.16

Consider the following system:

$$H(z) = \frac{z}{z^2 + 1}$$

What is $h(n)$, the impulse response?

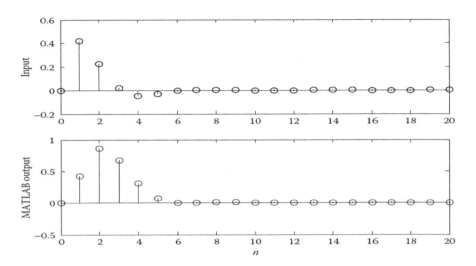

FIGURE 4.10 Signals for EOCE 4.15.

SOLUTION

We will use partial fraction expansion to find $h(n)$. As we did previously, we will put $H(z)/z$ in partial fraction form, then multiply by z to introduce a z into the numerator. By doing, so we will easily inverse-transform back to real time since the entries in Table 4.1 have z in the numerator.

$$\frac{H(z)}{z} = \frac{1}{z^2 + 1} = \frac{A}{z - j} + \frac{B}{z + j}$$

We next find the constants and write

$$\frac{H(z)}{z} = \frac{-1/2j}{z - j} + \frac{1/2j}{z + j}$$

Using Table 4.1, we get

$$h(n) = \frac{-1}{2} j(j)^n u(n) + \frac{1}{2} j(-j)^n u(n) = \frac{-1}{2} j\left(e^{j\pi/2n}\right) u(n) + \frac{1}{2} j\left(e^{-j\pi/2n}\right) u(n)$$

By taking common factors, we get

$$h(n) = \frac{1}{2} j\left[e^{-j\pi/2n} - e^{j\pi/2n}\right] u(n) = -\frac{1}{2} j\left[e^{j\pi/2n} - e^{-j\pi/2n}\right] u(n)$$

and finally

$$h(n) = -\frac{1}{2}(2j) j\sin\left(\frac{\pi}{2}n\right) u(n) = \sin\left(\frac{\pi}{2}n\right) u(n)$$

EOCE 4.17

Consider the system

$$H(z) = \frac{z}{z^2 - 2z + 2}$$

What is $h(n)$?

SOLUTION

The impulse response $h(n)$ is the inverse z-transform of $H(z)$. We will put $H(z)/z$ in partial fraction form first.

$$\frac{H(z)}{z} = \frac{1}{z^2 - 2z + 2} = \frac{A}{z - (1 + j)} + \frac{B}{z - (1 - j)}$$

And $H(z)$ will be

$$H(z) = \frac{zA}{z - (1 + j)} + \frac{Bz}{z - (1 - j)}$$

From this last expression for $H(z)$ and using Table 4.1, we get

$$h(n) = A\left(\sqrt{2}\right)^n e^{j\pi/4n}u(n) + B\left(\sqrt{2}\right)^2 e^{-j\pi/4n}u(n)$$

where $\sqrt{2}$ is the magnitude of $1 - j$ and $1 + j$ and $-\pi/4$ is the phase of $1 - j$.
   We still need to find A and B. We can use MATLAB to do that and to confirm some of the results that we arrived at (the magnitude and the phase of the poles). We will write the script EOCE4_17 to get

$$mag\_r = 0.5000 \ 0.5000$$

$$phase\_r = -1.5708 \ 1.5708$$

$$mag\_p = 1.4142 \ 1.4142$$

$$phase\_p = 0.7854 \ -0.7854$$

The impulse response is then

$$h(n) = 0.5e^{-1.57j}\left(\sqrt{2}\right)^n e^{j\pi/4n}u(n) + 0.5e^{1.57j}\left(\sqrt{2}\right)^n e^{-j\pi/4n}u(n)$$

In this expression, we have two complex conjugate terms. The sum of two conjugate terms is two times the real part of the term. Thus, we write

$$h(n) = 2\,\text{Real}\left[0.5e^{-1.57j}\left(\sqrt{2}\right)^n e^{\frac{j\pi}{4}n}\right] = \left(\sqrt{2}\right)^n \text{Real}\left[e^{-1.57j}e^{\frac{j\pi}{4}n}\right]$$

$$h(n) = \left(\sqrt{2}\right)^n \text{Real}\left[e^{j\left(\left(\frac{\pi}{4}\right)n - 1.57\right)}\right] = \left(\sqrt{2}\right)^n \cos\left(\frac{\pi}{4}n - 1.57\right)u(n)$$

EOCE 4.18

Consider the system

$$y(n) - .4y(n-1) = x(n)$$

with $x(n) = 2\sin\left(\frac{2\pi}{4}n\right)u(n)$. What are the initial and final values for $y(n)$ with $x(n)$ as the input?

SOLUTION

With

$$x(n) = 2\sin\left(\frac{2\pi}{4}n\right)u(n)$$

$X(z)$ is

$$X(z) = \frac{2z^{-1}\left(\sin(\pi/4)\right)}{1 - \left(2\cos\left((\pi/4)\right)\right)z^{-1} + z^{-2}} = \frac{2z\sin(\pi/4)}{z^2 - \left(2\cos(\pi/4)\right)z + 1} = \frac{\sqrt{2}}{z^2 - \sqrt{2}z + 1}$$

We know that $Y(z) = X(z)H(z)$. But from the difference equation, $H(z)$ is

$$H(z) = \frac{1}{1 - .4z^{-1}} = \frac{z}{z - .4}$$

Therefore,

$$Y(z) = \frac{z}{z - .4} \frac{\sqrt{2}z}{z^2 - \sqrt{2}z + 1}$$

The initial value of $y(n)$ is $y(0)$. Using the initial-value theorem, we can find $y(0)$ without solving for $y(n)$ from $Y(z)$:

$$y(0) = \lim_{z \to \infty} Y(z) = 0$$

The final value of $y(n)$ is

$$y(\infty) = \lim_{z \to 1}(z-1)\frac{z}{z-.4}\frac{\sqrt{2}z}{z^2 - \sqrt{2}z + 1}$$

provided that all poles of $(z - 1) Y(z)$ lie inside the unit circle. This is not the case here because the input is a pure sinusoid and it has its pole on the unit circle. Without paying attention to this, we may mistakenly use this theorem and write

$$y(\infty) = \lim_{z \to 1}(z-1)\frac{z}{z-.4}\frac{\left(\sqrt{2}/2\right)z}{z^2 - \sqrt{2}z + 1} = 0$$

But we know that the output $y(n)$ is also sinusoidal and does not settle on a single value as n approaches infinity. However, if $x(n) = u(n)$ then

$$Y(z) = \frac{z}{z-1}\frac{z}{z-.4}$$

and

$$y(0) = \lim_{z \to \infty}\frac{z^2}{(z-1)(z-.4)} = 1$$

The final value in this case will be

$$y(\infty) = \lim_{z \to 1}(z-1)\frac{z}{z-1}\frac{z}{z-0.4} = \frac{1}{1-.4} = \frac{1}{.6} = \frac{10}{6}$$

EOCE 4.19

Consider the system in Figure 4.11.

1. Find the transfer function representation.
2. Find the difference equation representation.
3. Find $y(n)$ if $x(n) = u(n)$.
4. Is the system stable?

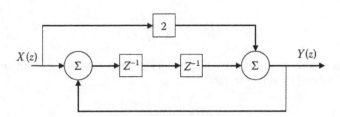

FIGURE 4.11   System for EOCE 4.19.

SOLUTION

1. The signal after the first summer is $X(z) + Y(z)$. This signal then passes through a delay and becomes $[X(z) + Y(z)]\, z^{-1}$. Then it passes through another delay and becomes $[[X(z) + Y(z)]\, z^{-1}]\, z^{-1}$. The output of the last summer is $Y(z)$ which is

$$Y(z) = 2X(z) + \left[\left[X(z) + Y(z)\right]z^{-1}\right]z^{-1} = 2X(z) + X(z)z^{-2} + Y(z)z^{-2}$$

By grouping like terms, we arrive at

$$Y(z)\left[1 - z^{-2}\right] = X(z)\left[2 + z^{-2}\right]$$

And the transfer function is

$$\frac{Y(z)}{X(z)} = H(z) = \frac{2 + z^{-2}}{1 - z^{-2}} = \frac{2z^2 + 1}{z^2 - 1}$$

2. From

$$\frac{Y(z)}{X(z)} = \frac{2 + z^{-2}}{1 - z^{-2}}$$

we can write

$$Y(z)\left[1 - z^{-2}\right] = X(z)\left[2 + z^{-2}\right]$$

The inverse transform is

$$y(n) - y(n-2) = 2x(n) + x(n-2)$$

3. If $x(n) = u(n)$ then $X(z) = z/(z-1)$. The output then is

$$Y(z) = \frac{z^2 + 2}{z^2 - 1}\frac{z}{z - 1}$$

$$\frac{Y(z)}{z} = \frac{z^2 + 2}{(z^2 - 1)(z - 1)} = \frac{z^2 + 2}{(z - 1)^2(z + 1)} = \frac{A}{z + 1} + \frac{B}{(z - 1)^2} + \frac{C}{z - 1}$$

The constants are found next.

$$A = \left.\frac{z^2 + 2}{(z - 1)^2}\right|_{z=-1} = \frac{1 + 2}{4} = \frac{3}{4}$$

$$B = \left.\frac{z^2 + 2}{z + 1}\right|_{z=1} = \frac{3}{2}$$

$$C = \frac{d}{dz}(B(z)) = \left.\frac{2z(z + 1) - (z^2 + 2)}{(z + 1)^2}\right|_{z=1} = \frac{4 - 3}{4} = \frac{1}{4}$$

Therefore, the output in the z-domain is

$$Y(z) = \frac{(3/4)z}{z+1} + \frac{(3/2)z}{(z-1)^2} + \frac{(1/4)z}{z-1}$$

The output in the time domain is

$$y(n) = \frac{3}{4}(-1)^n u(n) + \frac{3}{2}(1)^n \, nu(n) + \frac{1}{4}(1)^n u(n)$$

This solution can be verified using MATLAB as in the script EOCE4_19. The plots are in Figure 4.12.

4. To check stability, we look at the poles of

$$H(z) = \frac{z^2 + 2}{z^2 - 1} = \frac{z^2 + 2}{(z-1)(z+1)}$$

We can see that the poles are at $z = 1$ and $z = -1$ and they are on the unit circle. Hence, the system is on the verge of stability.

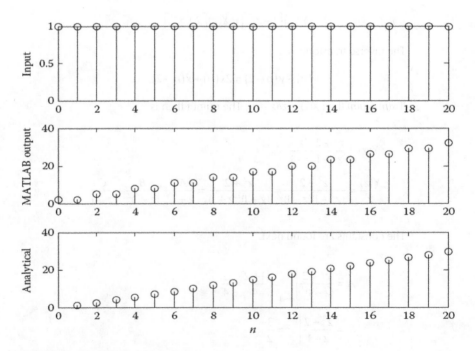

FIGURE 4.12   Signals for EOCE 4.19.

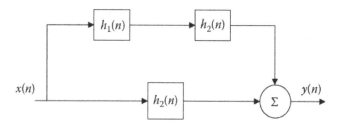

FIGURE 4.13 Signals for EOCE 4.20.

EOCE 4.20

Consider the system in Figure 4.13, with

$$H_1(z) = \frac{z}{z-.5} = H_2(z)$$

1. Find the transfer function.
2. Is the system stable?
3. Find the general form of $h(n)$ if $h(n)$ is real and causal.
4. If $x(n) = u(n)$, what is $y(n)$?

SOLUTION

1. From the Figure 4.13 we see that the output is

$$y(n) = \left[x(n) * h_1(n) \right] * h_2(n) + x(n) * h_2(n)$$

and from the properties of the z-transform we write Y(z) as

$$Y(z) = X(z)H_1(z)H_2(z) + X(z)H_2(z) = X(z)\left[H_1(z)H_2(z) + H_2(z) \right]$$

and the transfer function is

$$\frac{Y(z)}{X(z)} = H(z) = H_1(z)H_2(z) + H_2(z)$$

By substituting the individual transfer functions given to us, we arrive at

$$H(z) = \frac{z}{z-.5}\frac{z}{z-.5} + \frac{z}{z-.5} = \frac{z^2 + z(z-.5)}{(z-.5)^2} = \frac{2z^2 - .5z}{(z-.5)^2} = \frac{z(2z-.5)}{(z-.5)^2}$$

2. The poles are at $z = .5$ and they are within the unit circle. This means that the system is stable.

3. Since we are assuming that $h(n)$ is real and causal,

$$\frac{H(z)}{z} = \frac{2z - .5}{(z - .5)^2} = \frac{c_1}{(z - .5)^2} + \frac{c_2}{z - .5}$$

and

$$H(z) = \frac{zc_1}{(z - .5)^2} + \frac{zc_2}{z - .5}$$

With the help of Table 4.1, we get

$$h(n) = c_1 n (.5)^n u(n) + c_2 (.5)^n u(n)$$

This means that the ROC for $H(z)$ is $|z| > .5$. If the ROC for $H(z)$ were $|z| < .5$, then

$$h(n) = c_3 n (.5)^n (-n - 1) + c_4 (.5)^n u(-n - 1)$$

4. If the input $x(n) = u(n)$, then $X(z) = z/(z - 1)$ and

$$Y(z) = [H_1(z)H_2(z) + H_2(z)] \left[\frac{z}{z - 1}\right]$$

$$Y(z) = \frac{z}{(z - .5)^2} \frac{z}{z - 1} + \frac{z}{z - .5} \frac{z}{z - 1}$$

Let us write the output as the sum

$$Y(z) = Y_1(z) + Y_2(z)$$

and perform partial fraction expansion as

$$\frac{Y_1(z)}{z} = \frac{z}{(z - .5)^2(z - 1)} = \frac{c_1}{(z - .5)^2} + \frac{c_2}{z - .5} + \frac{c_3}{z - 1}$$

$$\frac{Y_2(z)}{z} = \frac{z}{(z - .5)(z - 1)} = \frac{c_4}{z - .5} + \frac{c_5}{z - 1}$$

Next, we find the constants as

$$c_1 = \left.\frac{z}{z - 1}\right|_{1/2} = \frac{1/2}{1/2 - 1} = -1$$

$$c_2 = \left.\frac{d}{dz} c_1(z)\right|_{z=1/2} = \left.\frac{z - 1 - z}{(z - 1)^2}\right|_{1/2} = \frac{-1}{1/4} = -4$$

$$c_3 = \left.\frac{z}{(z - .5)^2}\right|_{z=1} = \frac{1}{1/4} = 4$$

and

$$C_4 = \frac{z}{z-1}\Big|_{z=\frac{1}{2}} = \frac{\frac{1}{2}}{-\frac{1}{2}} = -1$$

$$C_5 = \frac{z}{z-.5}\Big|_{z=1} = \frac{1}{1-.5} = 2$$

The output in the z-domain is then

$$Y(z) = \frac{-1z}{(z-.5)^2} - \frac{4z}{z-.5} + \frac{4z}{z-1} - \frac{1z}{z-0.5} + \frac{2z}{z-1}$$

The output in the time domain is

$$y(n) = -n(.5)^n u(n) - 4(.5)^n u(n) + 4u(n) - (.5)^4 u(n) + 2u(n)$$

EOCE 4.21

If $x(n) = \delta(n) + \delta(n-3)$ and the output $y(n)$ is $\delta(n-4)$, what is the transfer function $H(z)$? Assume the system is linear and time invariant.

SOLUTION

We know that $y(n) = x(n) * h(n)$ in the discrete time domain. Mathematically,

$$y(n) = \sum_{m=-\infty}^{+\infty} x(m)h(n-m)$$

An easier way of finding $H(z)$ is to work directly with the z-domain by taking advantage of the convolution property of the z-transform

$$Y(z) = H(z)X(z)$$

After substitution, we get

$$z^{-4} = H(z)\left[1 + z^{-3}\right]$$

and finally

$$H(z) = \frac{z^{-4}}{1+z^{-3}} = \frac{1}{z^4 + z} = \frac{1}{z(z^3 + 1)}$$

4.15   END-OF-CHAPTER PROBLEMS

EOCP 4.1

Find the z-transform for the following signals and indicate the ROC.

1. $x(n) = 3(.3)^n u(n)$
2. $x(n) = (.3)^n u(n) - (.3)^n u(n-1)$
3. $x(n) = u(n) - u(n-1)$
4. $x(n) = \sin\left(n\dfrac{\pi}{3}\right)u(n) + (.3)^{n-1}u(n-1)$
5. $x(n) = u(n) * (.5)^n u(n)$
6. $x(n) = u(n) * (.5)^n u(n) * (.5)^{n-1} u(n-1)$
7. $x(n) = nu(n) - n\sin\left(\dfrac{2\pi}{3}n\right)u(n)$
8. $x(n) = (n-1)\,u(n-1) - 2\delta(n-1)$
9. $x(n) = u(-n-1) * u(n) + (n-1)\sin\left((n-1)\dfrac{\pi}{4}\right)u(n-1)$
10. $x(n) = n(.5)^n \sin(n)\,u(n) + u(-n-1)$

EOCP 4.2

Consider the following signals in the z-transform domain. Find $h(n)$ for each case.

1. $H(z) = 10\dfrac{z}{z-.5}$ $ROC : |z| < .5$
2. $H(z) = \dfrac{z^2 + z + 2}{(z-3)(z+3)(z-.1)}$ $ROC : .1 < |z| < 3$
3. $H(z) = \dfrac{z+1}{(z-.5)(z-.5)}$ $ROC : |z| > .5$
4. $H(z) = \dfrac{z+1}{(z-.5)^2(z-.3)}$ $ROC : .3 < |z| < .5$
5. $H(z) = \dfrac{(z-1)(z+1)}{z}$ $ROC : |z| > 0$

EOCP 4.3

The following signals will produce a causal $h(n)$. Find $h(n)$ using partial fraction, long division, and MATLAB.

1. $H(z) = \dfrac{1}{z(z-1)}$
2. $H(z) = \dfrac{1}{z(z-1)}$

3. $H(z) = \dfrac{z^2 + z + 1}{z^2 + 5z + 6}$

4. $H(z) = \dfrac{z + 1}{z^2 + 2z + 4}$

5. $H(z) = \dfrac{z^3 + z^2 + z + 1}{z}$

6. $H(z) = \dfrac{z^2 + 1}{z^3 + 2z^2 + 4z}$

EOCP 4.4

Draw the block diagrams for the following systems in the z-domain.

1. $\dfrac{Y(z)}{X(z)} = \dfrac{1}{z^2(z - 1)}$

2. $\dfrac{Y(z)}{X(z)} = \dfrac{z^2 + z}{z^2 + 5z + 6}$

3. $\dfrac{Y(z)}{X(z)} = \dfrac{z^2 + z + 1}{z^2 + 2z + 2}$

4. $\dfrac{Y(z)}{X(z)} = \dfrac{z^2 + 2z + 1}{z}$

5. $\dfrac{Y(z)}{X(z)} = \dfrac{z}{z^2 + 2z + 1}$

EOCP 4.5

Find $Y(z)/X(z)$ for the block diagrams in Figures 4.14–4.18.

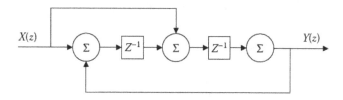

FIGURE 4.14 Block for EOCP 4.5.

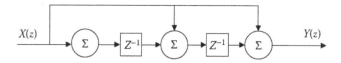

FIGURE 4.15 Block for EOCP 4.5.

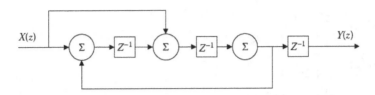

FIGURE 4.16   Block for EOCP 4.5.

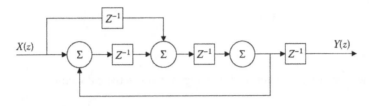

FIGURE 4.17   Block for EOCP 4.5.

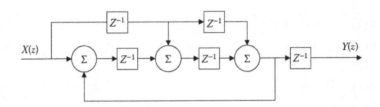

FIGURE 4.18   Block for EOCP 4.5.

EOCP 4.6

For each block diagram in Figures 4.19–4.23, find $y(n)$, the step response.

FIGURE 4.19   Block for EOCP 4.6.

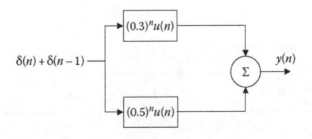

FIGURE 4.20   Block for EOCP 4.6.

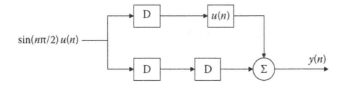

FIGURE 4.21 Block for EOCP 4.6.

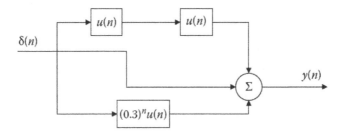

FIGURE 4.22 Block for EOCP 4.6.

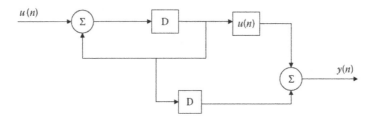

FIGURE 4.23 Block for EOCP 4.6.

EOCP 4.7

Consider the following transfer functions. Find the difference equations representing these systems and indicate if any of them is stable. Use MATLAB to find $h(n)$.

1. $H(z) = \dfrac{z+2}{z^2 + 2z + 5}$

2. $H(z) = \dfrac{z+2}{z^3 + 2z^2 + 5z}$

3. $H(z) = \dfrac{1}{(2z^3 - 1)(z - .1)}$

4. $H(z) = \dfrac{z^2 + z + 1}{(z^2 + .2)z}$

5. $H(z) = \dfrac{1}{z^3 + 3z^2 + 2z + 1}$

EOCP 4.8

Consider the following difference equations. Find the transfer functions and the outputs $y(n)$. Are the systems stable?

1. $y(n) - .5y(n-1) = x(n) + x(n-1)$, $y(-1) = 0$, $x(n) = \delta(n)$
2. $y(n) - .5y(n-1) = x(n) + x(n-1)$, $y(-1) = 1$, $x(n) = \delta(n)$
3. $y(n) - .5y(n-1) = x(n) + x(n-1)$, $y(-1) = 0$, $x(n) = \sin(2\pi/1n)u(n)$
4. $y(n) - .3y(n-1) = x(n) + x(n-1)$, $y(-1) = -1$, $x(n) = \sin(2\pi n/3)\,u(n) + \cos(2\pi n/3)u(n)$
5. $y(n) - .8y(n-1) + . 2y(n-2) = x(n)$, $y(-1) = 0$, $y(-2) = 1$, $x(n) = \delta(n)$
6. $y(n) - .8y(n-1) + . 2y(n-2) = x(n-1)$, $y(-1) = 1$, $y(-2) = 1$, $x(n) = u(n)$
7. $y(n) + y(n-1) + y(n-2) = x(n)$, $y(-1) = y(-2) = 0$, $x(n) = u(n)$
8. $y(n) + .1y(n-2) = x(n-2)$, $y(-1) = y(-2) = 1$, $x(n) = u(n)$
9. $y(n) - .5y(n-3) = x(n)$, $y(-1) = y(-2) = 0$, $y(-3) = 1$, $x(n) = \delta(n)$
10. $y(n) = x(n) + x(n-1) + x(n-2) + x(n-3)$, $x(n) = \sin(n)u(n)$

EOCP 4.9

The following are outputs of linear discrete systems. Find $y(0)$ and $y(\infty)$ for each output.

1. $Y(z) = \dfrac{z}{z-10}$

2. $Y(z) = \dfrac{z^2 + z}{z^2 + 2z + 1}$

3. $Y(z) = \dfrac{z^2 + 2z + 1}{(z-1)(z+2)}$

4. $Y(z) = \dfrac{z^2 + 3z + 1}{(z-.1)(z-.2)(z-.3)}$

5. $Y(z) = \dfrac{z}{z-1} + \dfrac{z}{(z-1)^2}$

EOCP 4.10

Find the steady-state output for the systems:

1. $H(z) = \dfrac{1}{z-.5} x(n) = \sin(n)u(n) + \cos(n)u(n)$

2. $H(z) = \dfrac{1}{(z-.2)(z-.3)}$ $x(n) = \sin\left(\dfrac{2\pi}{3}n\right)u(n)$

3. $H(z) = \dfrac{z+1}{z} x(n) = 10$

4. $H(z) = \dfrac{z+1}{z^2}\, x(n) = \delta(n) + u(n)$

5. $H(z) = \dfrac{z^2 + z + 1}{z}\, x(n) = 3$

EOCP 4.11

Consider the following difference equation:

$$y(n) - .5y(n-1) - .3y(n-2) - .4y(n-3) = x(n)$$

1. With $y(-1) = y(-2) = y(-3) = 0$ and
   a. $x(n) = \delta(n)$
   b. $x(n) = u(n)$
   c. $x(n) = \cos(10\pi n/3)u(n)$

Use MATLAB to find $y(n)$ assuming that $y(n)$ is causal; use long division to verify the MATLAB results.

2. Is the system stable?
3. Find $h(n)$ if ROC is $|z| > 2$.

EOCP 4.12

Consider the system in Figure 4.24.

1. Find the difference equation representing the system.
2. For what value(s) of a is the system stable?
3. Pick a value for a to make the system stable, and find $y(n)$ for $x(n) = \delta(n)$
4. Use MATLAB to find $y(n)$ for $a = .5$ if
   a. $x(n) = 10\,\delta(n)$
   b. $x(n) = 10\sin(3n/4\pi)u(n)$
5. Find the initial and final values for $y(n)$ in part 4.

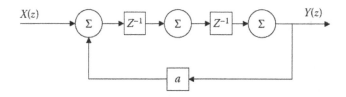

FIGURE 4.24   Block for EOCP 4.12.

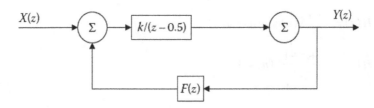

FIGURE 4.25   Block for EOCP 4.13.

EOCP 4.13

Consider the following system in Figure 4.25.

1. Find $Y(z)/X(z)$ as a function of $F(z)$, the feedback function.
2. If $F(z) = 1$, find $Y(z)/X(z)$.
3. Is the system stable in 2? If yes, for what k values?
4. If $F(z) = z/(z - .2)$, find $Y(z)/X(z)$.
5. Is the system stable in 4? If yes, for what k values?
6. Find the difference equation representing the system for both cases

$$F(z) = 1 \text{ and } F(z) = z/(z - .2).$$

7. Find the impulse response of the system when $F(z) = 1$ and $F(z) = z/(z - .2)$ with a value of k that makes the system stable.
8. Find the step response for both $F(z) = 1$ and $F(z) = z/(z - .2)$ with a k value that stabilizes the system.
9. Let $E(z) = X(z) - F(z) \, Y(z)$. Find this error signal $E(z)$ as a function of k.
10. Find $e(0)$ and $e(\infty)$. $e(n)$ is called the error signal.
11. What value for k will make $e(\infty) = .1$? Pick a suitable input $x(n)$ that will give you this k value.

EOCP 4.14

The outputs and the inputs are given for an unknown linear time-invariant system. What is the transfer function for each?

1. $x(n) = \delta(n) + \delta(n - 1) + \delta(n - 2)$
   $y(n) = \delta(n - 1) + \delta(n - 2) + \delta(n - 3)$
2. $x(n) = \sin\left(\frac{2\pi}{3}n\right)u(n)$

   $y(n) = \cos\left(\frac{2\pi}{3}n\right)u(n)$
3. $x(n) = u(n)$
   $y(n) = nu(n)$

4. $x(n) = \delta(n)$
   $y(n) = u(n)$

5. $x(n) = 2\sin\left(\dfrac{3\pi}{11}n\right)u(n) + u(n)$

   $y(n) = 10\cos\left(\dfrac{3\pi}{11}n\right)u(n)$

5 Discrete Fourier Transform and Discrete Systems

5.1 INTRODUCTION

We have two types of linear systems: the continuous linear system and the discrete linear system. The Fourier transform of the continuous signal $x(t)$ is

$$X(w) = \int_{-\infty}^{+\infty} x(t) e^{-jwt} dt \qquad (5.1)$$

By observing the signal $x(t)$ in the frequency domain, we are able to see all the frequencies that $x(t)$ contains, and based on that we can further process the signal $x(t)$ and alter its frequency components in any way we wish.

But the Fourier transform Equation (5.1) is not suitable for computer evaluation since the bounds on the integral are infinite. By *discretizing* the signal $x(t)$ and obtaining the discrete signal $x(n)$, we can Fourier transform the signal $x(n)$ into the frequency domain in order to see its frequency components hoping that we can utilize the computer in all calculations. The Fourier transform of the signal $x(n)$ is

$$X\left(e^{j\theta}\right) = \sum_{-\infty}^{+\infty} x(n) e^{-j\theta n} \qquad (5.2)$$

where θ is the digital frequency and is related to the continuous frequency by the equation

$$\theta = wT_s \qquad (5.3)$$

where T_s is the sampling period used to sample $x(t)$ to get $x(n)$. $x(n)$ now is discrete but is of infinite duration and, again, it is not possible to evaluate $X(e^{j\theta})$ on the computer. This problem will be discussed next.

5.2 DISCRETE FOURIER TRANSFORM AND THE FINITE-DURATION DISCRETE SIGNALS

Continuing with the earlier discussion, let us assume that $x(n)$ is zero for all the integer n values less than zero, and for n greater than or equal to some integer value N. In other words, suppose $x(n) = 0$ for $n < 0$ and $n \geq N$. Then we can define the N-points Fourier transform for the truncated $x(n)$ as

$$X(k) = \sum_{n=0}^{N-1} x(n) e^{-j2\pi nk/N} \qquad k = 0, 1, \ldots, N-1 \qquad (5.4)$$

Equation (5.4) is the discrete Fourier transform (DFT) of the finite duration signal $x(n)$.

In this defining equation,

1. $x(n)$ in the truncated signal that is zero for $n < 0$ and $n \geq N$.
2. N is the length of the truncated signal $x(n)$.
3. n is the index for the samples in $x(n)$.
4. k is the frequency index for the DFT.

We can see in Equation (5.4) for the DFT that $X(k)$ is finite, $x(n)$ is finite, and the number of points in $x(n)$ is the same as the number of points in $X(k)$. Now it is easy to implement the DFT equation using the computer.

To go back to the discrete time signal $x(n)$ from the $X(k)$ values, we use the inverse relation

$$x(n) = \frac{1}{N} \sum_{k=0}^{N-1} X(k) e^{-j2\pi nk/N} \qquad n = 0, 1, \ldots, N-1 \qquad (5.5)$$

Example 5.1

Consider the discrete signal given by

$$x(0) = 1 \text{ and } x(1) = 0$$

Find the DFT of $x(n)$ and do the inverse DFT to obtain $x(n)$ from $X(k)$ for $n = 0, 1$.

SOLUTION

In this example, N is 2 and $X(k)$ is

$$X(k) = \sum_{n=0}^{1} x(n) e^{-j2\pi nk/N} = \sum_{n=0}^{1} x(n)(W_n)^{nk}$$

Before we evaluate $X(k)$, let us look at the defining equation for $X(k)$. We have substituted $W_N = e^{-j2\pi/N}$ in Equation (5.5) where

$$W_N^{nk} = e^{-j2\pi nk/N} = (W_N)^{nk}$$

Based on this relation, in this example we have

$$W_N = W_2 = e^{-j2\pi/2} = e^{-j\pi} = \cos(\pi) - j\sin(\pi) = -1$$

Now by expanding the $X(k)$ equation, we get

$$X(0) = x(0)(-1)^0 + x(1)(-1)^0 = 1$$

$$X(1) = x(0)(-1)^0 + x(1)(-1)^1 = 1$$

To calculate $x(n)$ from $X(k)$, we use the inverse relation

$$x(n) = \frac{1}{N}\sum_{k=0}^{1} X(k)e^{j2\pi nk/N}$$

to get

$$x(0) = \frac{1}{2}X(0)e^{j\pi 0} + \frac{1}{2}X(1)e^{j\pi 0} = \frac{1}{2}(1) + \frac{1}{2}(1) = 1$$

and

$$x(1) = \frac{1}{2}X(0)e^{j\pi 0} + \frac{1}{2}X(1)e^{j\pi} = \frac{1}{2}(1) + \frac{1}{2}(1)(-1) = 0$$

5.3 PROPERTIES OF THE DFT

Following are some of the characteristics that the DFT possesses (Table 5.1).

TABLE 5.1
Some Properties of the DFT

Linearity	$\alpha x_1(n) + \beta x_2(n) \leftrightarrow \alpha X_1(k) + \beta X_2(k)$
Time shifting	$x(n-m) \leftrightarrow X(k)e^{-j2\pi km/N}$
Frequency shifting	$x(n)e^{j2\pi nm/N} \leftrightarrow X(k-m)$
Modulation	$x_1(n)x_2(n) \leftrightarrow \dfrac{1}{N} X_1(k) \otimes X_2(k)$
Circular convolution	$x_1(n) \otimes x_2(n) \leftrightarrow X_1(k)X_2(k)$
Parseval's theorem	$\displaystyle\sum_{k=0}^{N-1}\lvert x(n)\rvert^2 = \frac{1}{N}\sum_{k=0}^{N-1}\lvert X(k)\rvert^2$
Duality	$\dfrac{1}{N}X(n) \leftrightarrow x(-k)$

$\otimes$ indicates circular convolution.

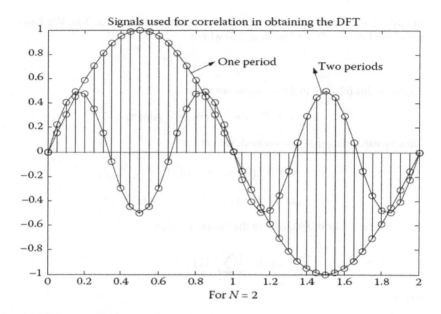

FIGURE 5.1 Correlation signals for the DFT.

5.3.1 How Does the Defining Equation Work?

If we look at the defining equation in its trigonometric form, we have

$$X(k) = \sum_{n=0}^{N-1} x(n) \left[\cos\left(\frac{2\pi nk}{N}\right) - j\sin\left(\frac{2\pi nk}{N}\right) \right] \tag{5.6}$$

You can see that each component of $X(k)$ is a correlation of all the samples in $x(n)$ with cosine and sine signals that have k complete periods in the N samples interval. This can be seen in the example that we considered earlier, Example 5.1, where with $N = 2$

$$X(k) = \sum_{n=0}^{1} x(n) \left[\cos(\pi nk) - j\sin(\pi nk) \right]$$

If we plot $\cos(\pi nk)$ or $\sin(\pi nk)$ on a fine grid, we will observe a complete cycle for $k = 1$ and 2 cycles for $k = 2$ in the $N = 2$ interval. This can be seen in Figure 5.1.

5.3.2 DFT Symmetry

For a real $x(n)$ signal, the DFT is symmetrical around the point $k = N/2$. The magnitude of $X(k)$ will have even symmetry and the phase of $X(k)$ will have odd symmetry. To see this, we can show that

$$|X(k)| = |X(N - k)|$$

and that the phase of $X(k)$ is the negative of the phase of $X(N - k)$. The magnitude of $X(k)$ is

$$\left| X(k) \right| = \left| \sum_{n=0}^{N-1} x(n) e^{-j2\pi n/N} \right|$$

and the magnitude of $X(N - k)$ is

$$\left| X(N - k) \right| = \left| \sum_{n=0}^{N-1} x(n) e^{-j2\pi n(N-k)/N} \right| = \left| \sum_{n=0}^{N-1} x(n) e^{j2\pi n/N} \right|$$

since $e^{-j2\pi n N/N} = 1$.

To look at this in a different way, we have

$$X(K) = \sum_{n=0}^{N-1} x(n) \left[\cos\left(\frac{2\pi n}{N} \right) - j\sin\left(\frac{2\pi n}{N} \right) \right]$$

But

$$\left| X(k) \right| = \left| X(N - k) \right| \quad \text{if Real}\left[X(k) \right] = \text{Real}\left[X(N - k) \right]$$

$$\text{Real}\left[X(k) \right] = \sum_{n=0}^{N-1} x(n) \cos\left(\frac{2\pi n}{N} \right)$$

and

$$\text{Real}\left[X(N - k) \right] = \sum_{n=0}^{N-1} x(n) \cos\left(\frac{2\pi n(N - k)}{N} \right)$$

We also know that

$$\cos\left(2\pi n - \frac{2\pi n}{N} \right) = \cos\left(\frac{-2\pi n}{N} \right) = \cos\left(\frac{2\pi n}{N} \right)$$

Thus,

$$\text{Real}\left[X(k) \right] = \text{Real}\left[X(N - k) \right]$$

and consequently

$$\left| X(k) \right| = \left| X(N - k) \right|$$

To show that

$$\angle X(k) = -\angle X(N - K)$$

we need to show that the imaginary

$$\text{Imag}\big[X(k)\big] = -\text{Imag}\big[(X(N-K))\big]$$

But

$$\text{Imag}\big[X(k)\big] = -\sum_{n=0}^{N-1} x(n)\sin\left(\frac{2\pi nk}{N}\right)$$

and

$$\text{Imag}\big[X(N-k)\big] = -\sum_{n=0}^{N-1} x(n)\sin\left(2\pi n\frac{(N-k)}{N}\right)$$

But again

$$\sin\left(2\pi n + -\frac{2\pi nk}{N}\right) = \sin\left(\frac{-2\pi nk}{N}\right) = -\sin\left(\frac{2\pi nk}{N}\right)$$

Thus,

$$\text{Imag}\big[X(N-k)\big] = \sum_{n=0}^{N-1} x(n)\sin\left(\frac{2\pi nk}{N}\right)$$

Therefore,

$$\text{Imag}\big[X(k)\big] = -\text{Imag}\big[(X(N-k))\big]$$

and thus

$$\angle X(k) = -\angle X(N-k)$$

5.3.3 DFT LINEARITY

Given the signal

$$X(n) = \alpha x_1(n) + \beta x_2(n)$$

the DFT of $x(n)$ is then

$$X(k) = \alpha X_1(k) + \beta X_2(k)$$

To see this result, we have

$$X(k) = \sum_{n=0}^{N-1}\big(\alpha x_1(n) + \beta x_2(n)\big)e^{-j2\pi nk/N} = \alpha\sum_{n=0}^{N-1} x_1(n)e^{-j2\pi nk/N} + \beta\sum_{n=0}^{N-1} x_2(n)e^{-j2\pi nk/N}$$

Finally, we have

$$X(k) = \alpha X_1(k) + \beta X_2(k)$$

This linearity property is very important because we usually have signals that have more than one frequency component.

5.3.4 MAGNITUDE OF THE DFT

If we are calculating the DFT of the signal $x(n)$ that was obtained by taking N samples of a real sinusoid of amplitude M_r, then the magnitude of the DFT is

$$|X(k)| = \frac{M_r N}{2}$$

If the N samples in $x(n)$ were obtained by sampling a complex sinusoid of the form

$$x(t) = M_c[\cos(wt) + j\sin(wt)] = M_c e^{jwt}$$

then

$$|X(k)| = M_c N$$

If, however, $x(n)$ has a dc component M_d, then

$$|X(k)| = M_d N$$

In this relation, the input to the DFT system should have a sinusoidal component that makes an integer number of periods over the N samples.

5.3.5 WHAT DOES k IN $X(k)$, THE DFT, MEAN?

After we calculate the DFT of the signal $x(n)$, which is usually sample of the continuous signal $x(t)$, we may be interested in the value of the frequency when the magnitude of $X(k)$ is maximum. The plot of $|X(k)|$ is usually drawn versus the frequency index k. The distance between the successive values of k is given by the frequency resolution f_s/N, where f_s is the sampling frequency. If the maximum value of $|X(k)|$ is at $k = k_m$, for example, then the frequency at which $|X(k)|$ is maximum is $k_m f_s/N$ Hz. If we let $\Delta f = f_s/N$ be the frequency resolution, then we have

$$\Delta f = \frac{f_s}{N} = \frac{1}{NT_s} = \frac{1}{T_r} \tag{5.7}$$

where T_r is the time interval along which the samples are taken.

Example 5.2

Consider the magnitude DFT of the signal $x(n)$ that was obtained by sampling a continuous sinusoid at $f_s = 1000$ Hz, as shown in Figure 5.2.

1. What is the frequency at which $|X(k)|$ is maximum?
2. What is the frequency at which $|X(k)|$ is minimum?
3. If only the value at $k = 2$ is known for $|X(k)|$, can you find $|X(3)|$?

SOLUTION

1. From the plot we see that $|X(k)|$ is maximum at $k = 0$ and $k = 4$ and that $N = 4$. Thus, we have $k_m f_s/N = 0(1000)/4 = 0$ Hz and $4(1000)/4 = 100$ Hz as the two frequencies.
2. The minimum magnitude is at $k = 1$, 2, and 3. Thus,

$$\frac{1(1000)}{4} = 250 \qquad \frac{2(1000)}{4} = 500 \qquad \frac{3(1000)}{4} = 750 \text{ Hz}$$

are the frequencies.
3. With only $X(2)$ known, we have

$$\frac{N}{2} = \frac{4}{2} = 2$$

Thus, from the symmetry property, we would not know $X(3)$. There is no need to use the value at $k = 4$ since the magnitude is periodic.

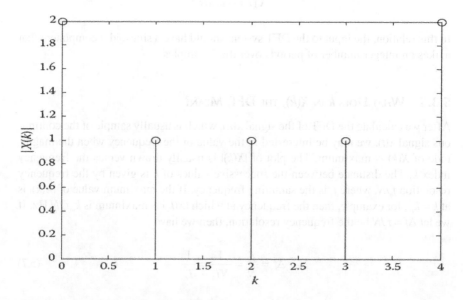

FIGURE 5.2 Signal for Example 5.2.

5.4   RELATION THE DFT HAS WITH THE FOURIER TRANSFORM OF DISCRETE SIGNALS, THE z-TRANSFORM, AND THE CONTINUOUS FOURIER TRANSFORM

5.4.1   DFT AND THE FOURIER TRANSFORM OF x(n)

Given $x(n)$ for $-\infty \leq n \leq +\infty$. The Fourier transform is defined as

$$X\left(e^{j\theta}\right) = \sum_{n=-\infty}^{+\infty} x(n) e^{-j\theta n}$$

where $\theta = wT_s$, w is the continuous radian frequency and T_s is the sampling interval used to sample $x(t)$ to get $x(n)$. If the signal $x(n)$ is truncated and is limited to the lower $n = 0$ and the upper $n = N - 1$ interval, where N is some integer value, then

$$X\left(e^{j\theta}\right) = \sum_{n=0}^{N-1} x(n) e^{-j\theta n}$$

Now if we let

$$\theta = \frac{2\pi k}{N}$$

we will have

$$X\left(e^{j2\pi k/N}\right) = \sum_{n=0}^{N-1} x(n) e^{-j2\pi kn/N}$$

But the term $2\pi k/N$ is a function of k only since N is known. Thus, we can write

$$X(k) = \sum_{n=0}^{N-1} x(n) e^{-j2\pi kn/N} = X\left(e^{j\theta}\right)\Big|_{\theta=2\pi k/N} \qquad (5.8)$$

Equation (5.8) is nothing but the DFT of the finite signal $x(n)$ defined for $n = 0, 1, ..., N - 1$. Therefore, from this development you can see that if $x(n)$ is limited to the interval $n = 0, 1, ..., N - 1$, and $X(e^{j\theta})$ is sampled starting at values of $\theta = 0$ and then regularly spaced along the unit circle by $2\pi k/N$, the DFT will be the sampled version of $X(e^{j\theta})$.

5.4.2   DFT AND THE z-TRANSFORM OF x(n)

The z-transform for a finite length signal $x(n)$ is

$$X(z) = \sum_{n=0}^{N-1} x(n) z^{-n}$$

where N is the number of samples in $x(n)$. If we set $z = e^{j2\pi k/N}$, then

$$X\left(e^{j2\pi k/N}\right) = \sum_{n=0}^{N-1} x(n)e^{-j2\pi nk/N}$$

But again, the term $2\pi k/N$ is a function of k only since N is known. Thus, we can write

$$X(k) = \sum_{n=0}^{N-1} x(n)e^{-j2\pi nk/N} = X(z)\bigg|_{z=e^{j2\pi nk/N}} \qquad (5.9)$$

5.4.3 DFT AND THE CONTINUOUS FOURIER TRANSFORM OF $x(t)$

Consider the signal $x(t)$ that is assumed to be zero for $t < 0$. The Fourier transform of $x(t)$ is

$$X_c(w) = \int_0^\infty x(t)e^{-jwt}\,dt$$

Let T_s be a small positive number. Then the previous equation can be approximated as

$$X_c(w) = \sum_{n=0}^\infty \int_{nT_s}^{nT_s+T_s} x(t)e^{-jwt}\,dt = \int_0^{T_s} x(t)e^{-jwt}\,dt + \int_{T_s}^{2T_s} x(t)e^{-jwt}\,dt + \ldots$$

In the interval $nT_s \le t \le nT_s + T_s$, if T_s is chosen small enough such that $x(t)$ in this interval will be considered constant, then

$$X_c(w) = \sum_{n=0}^\infty \int_0^{T_s} e^{-jwt}\,dt\,x(nT_s)$$

With

$$\int_{nT_s}^{nT_s+T_s} e^{-jwt}\,dt = \frac{e^{-jwtT_s} - e^{-(1+n)jwT_s}}{jw}$$

we have

$$X_c(w) = \frac{1-e^{-jwT_s}}{jw} \sum_{n=0}^\infty e^{-jwnT_s} x(nT_s)$$

If $x(nT_s)$ is very small for $n \geq N$, then the continuous Fourier transform can be simplified further as

$$X_c(w) = \frac{1-e^{-jwT_s}}{jw} \sum_{n=0}^{N-1} e^{-jwnT_s} x(nT_s)$$

If we set $w = 2\pi k/NT_s$, we will have

$$X_c\left(\frac{2\pi k}{NT_s}\right) = \frac{1-e^{-jwT_s}}{j(2\pi k/NT_s)} \sum_{n=0}^{N-1} e^{-j2\pi kn/N} x(nT_s)$$

But

$$\sum_{n=0}^{N-1} e^{-j2\pi nk/N} x(nT_s) = X(k)$$

which is the DFT of $x(nT_s)$. Thus, the continuous Fourier transform can be approximated at $(2\pi/NT_s)k$ points, where $k = 0, 1, \ldots, N-1$. Finally, we write

$$X_c\left(\frac{2\pi k}{NT_s}\right) = \frac{1-e^{-jwT_s}}{j(2\pi k/NT_s)} X(k) \qquad (5.10)$$

Example 5.3

Consider the signal $x(n)$, where $x(0) = 1$, $x(1) = 0$, and $x(2) = 1$. Use the z-transform method to find $X(k)$.

SOLUTION

This signal can easily be put in the z-domain. It is

$$X(z) = 1z^{-0} + 0z^{-1} + 1z^{-2} = 1 + z^{-2}$$

With $z = e^{j\theta}$, we have

$$X(e^{j\theta}) = 1 + e^{-2j\theta}$$

But

$$X(k) = X(e^{j\theta})\big|_{\theta=2\pi k/N}$$

With $N = 3$ as is the case in this example, we have

$$X(k) = 1 + e^{-2j(2\pi k/3)}$$

with

$$X(0) = 1 + e^{-2j(0)} = 2$$

$$X(1) = 1 + e^{-2j(2\pi/3)}$$

$$X(2) = 1 + e^{-2j(2\pi 2/3)} = 1 + e^{-j(8\pi/3)}$$

Using the defining equation of the DFT, we have

$$X(k) = \sum_{n=0}^{N-1} x(n) e^{-2j\pi nk/N} = \sum_{n=0}^{2} x(n)(W_N)^{nk}$$

with

$$e^{-j2\pi/N} = e^{-j\pi 2/3}$$

With this simplification, we can get the first few samples of the DFT.

$$X(0) = x(0) e^{-j2\pi/3(0)} + x(1) e^{-j2\pi/3(0)} + x(2) e^{-j2\pi/3(0)} = 2$$

$$X(1) = x(0) e^{-j2\pi/3(0)} + x(1) e^{-j2\pi/3(1)} + x(2) e^{-j2\pi/3(2)} = 1 + e^{-j4\pi/3}$$

$$X(2) = x(0) e^{-j2\pi/3(0)} + x(1) e^{-j2\pi/3(2)} + x(2) e^{-j2\pi/3(4)} = 1 + e^{-j8\pi/3}$$

5.5 NUMERICAL COMPUTATION OF THE DFT

The DFT is a computer program that can be used to transform the signal $x(n)$ defined for $n = 0, 1, ..., N - 1$, where N is the number of samples in $x(n)$, to $X(k)$, a set of N values defined for the frequency index $k = 0, 1, ..., N - 1$. In other words, the DFT can be thought of as a system (a computer program) whose input is $x(n)$ and whose output is $X(k)$.

To implement the DFT equation on a digital computer, MATLAB® will be used. To do that, we will try to find the DFT of the signal in Example 5.3, where $x(0) = 1$, $x(1) = 0$, and $x(2) = 1$. The MATLAB script XFromx_n is used.

And the results are the same as those obtained in Example 5.3, where $x(0) = 2$, $x(1) = 1 + e^{-j4\pi/3}$, and $x(2) = 1 + e^{-j8\pi/3}$. To find $x(n)$ from $X(k)$, we use the inverse DFT and write the MATLAB script x_nFromX.

The results are again similar to the original signal $x(n)$, with $x(0) = 1$, $x(1) = 0$, and $x(2) = 1$.

5.6   FAST FOURIER TRANSFORM: A FASTER WAY OF COMPUTING THE DFT

If we look at the following equation to calculate $X(k)$ from $x(n)$:

$$X(k) = \sum_{n=0}^{N-1} x(n) e^{-j2\pi nk/N}$$

we can see that to evaluate $X(k)$ for a single k value we need to perform N multiplications. Yet to calculate the N values for k we will need N^2 multiplications. The goal is to reduce this number of multiplications. For that reason, the fast Fourier transform (FFT), a fast way of computing the DFT, was developed in which the number of multiplications N^2 was reduced to $N(\log_2 N)/2$ multiplications. Thus, if $N = 1024$, for example, it will take 10,48,576 multiplications using the DFT. Using the FFT, it will take 5120 multiplications, a drastic reduction in the number of multiplications.

To see how this FFT was developed, let us look back at the DFT equation. We have

$$X(k) = \sum_{n=0}^{N-1} x(n) e^{-j2\pi nk/N}$$

If we define $W_N = e^{-j2\pi/N}$, then the DFT equation becomes

$$X(k) = \sum_{n=0}^{N-1} x(n) W_N^{nk}$$

Here we will derive what is called the radix-2 FFT and use what is called decimation in time. The last equation can be written as

$$X(k) = \sum_{n=0}^{(N/2)-1} x(2n) W_N^{2nk} + \sum_{n=0}^{(N/2)-1} x(2n+1) W_N^k W_N^{2nk}$$

With W_N^k not depending on the index n, this term can be pulled out and $X(k)$ becomes

$$X(k) = \sum_{n=0}^{(N/2)-1} x(2n) W_N^{2nk} + W_N^k \sum_{n=0}^{(N/2)-1} x(2n+1) W_N^{2nk}$$

Notice that we are dividing $x(n)$ into even and odd parts. Notice also that

$$W_{N/2} = e^{-j2\pi/(N/2)} = e^{-j2\pi 2/N} = W_N^2$$

With this observation, $X(k)$ becomes

$$X(k) = \sum_{n=0}^{(N/2)-1} x(2n)W_{N/2}^{2nk} + W_N^k \sum_{n=0}^{(N/2)-1} x(2n+1)W_{N/2}^{nk} \tag{5.11}$$

Now consider $X(k + (N/2))$. With $W_{N/2}^{n(k+(N/2))} = W_{N/2}^{nk}$ and $W_N^{k+(N/2)} = -W_N^k$ we can write $X(k + (N/2))$ as

$$X\left(k + \frac{N}{2}\right) = \sum_{n=0}^{(N/2)-1} x(2n)W_{N/2}^{nk} - W_N^k \sum_{n=0}^{(N/2)-1} x(2n+1)W_{N/2}^{2nk} \tag{5.12}$$

with $X(k)$ repeated as

$$X(k) = \sum_{n=0}^{(N/2)-1} x(2n)W_{N/2}^{2nk} + W_N^k \sum_{n=0}^{(N/2)-1} x(2n+1)W_{N/2}^{nk} \tag{5.13}$$

In Equation (5.13), we can find the first $N/2$ points. We can use Equation (5.12) for $X(k + N/2)$ to find the other $N/2$. Notice that the last two equations are similar with only the sign of W_N^k reversed.

To make things more clear, let us consider the case where $N = 8$, an 8-point DFT. With the last two equations, we can see that the 8-point DFT is reduced to the 4-point DFT. Remember that a 2-point DFT can be implemented without any complex multiplication. So if we subdivide the 4-point DFT further, we will get the 2-point DFT.

In general, if we are given N samples of $x(n)$ and are asked to find $X(k)$, we will first subdivide the N-point DFT into $N/2$-point DFT. We continue with this process until we get to the 2-point DFT.

The FFT is implemented in MATLAB and can be used directly to evaluate the DFT of the signal $x(n)$. We need to remind the reader here that the FFT will give the same result as the DFT. The FFT is not an approximation to the DFT in any way. Also, the properties for the DFT will hold for the FFT.

5.7   APPLICATIONS OF THE DFT

We will look next at some of the important applications of the DFT.

5.7.1   CIRCULAR CONVOLUTION

Let us look at the input–output relationship between the input $x(n)$ and the output $y(n)$ of the discrete linear system.

$$y(n) = x(n) * h(n) = \sum_{m=-\infty}^{\infty} x(m)h(n-m)$$

where * indicates linear convolution. If we define the signals $x(n)$ and $h(n)$ to have zero values for $n < 0$ and $n \geq N$, then the linear convolution relationship becomes

$$y(n) = \sum_{m=0}^{2N-1} x(m)h(n-m)$$

But we can see that if $n < 0$ and $n > 2N - 1$, $h(n - m)$ is zero. Hence

$$y(n) = 0 \quad \text{for } n < 0 \text{ and } n > 2N - 1$$

Notice that $2N - 1$ is the length of the convolution result. However, $y(n)$ is generally not zero for $n \geq N$. Note also that if we take the N-points DFT of $y(n)$, we will miss the nonzero values of $y(n)$ as the result of the linear convolution. This indicates a need to define another form of convolution so that $y(n) = x(n) * h(n)$ is zero for $n < 0$ and $n \geq N$.

To arrive at this conclusion in a different approach, let $X(k)$, $H(k)$, and $Y(k)$ be the DFTs of the input $x(n)$, the impulse response $h(n)$, and the output $y(n)$ of the linear discrete system.

Let us claim now that

$$Y(k) = X(k)H(k) \tag{5.14}$$

Does this mean that

$$y(n) = x(n) * h(n)$$

where * indicates convolution.

Taking the inverse DFT of $Y(k) = X(k)H(k)$ leads to

$$y(n) = \frac{1}{N} \sum_{k=0}^{N-1} \left[\sum_{m=0}^{N-1} x(m)e^{-j2\pi mk/N} \sum_{p=0}^{N-1} h(p)e^{-j2\pi pk/N} \right] e^{j2\pi nk/N}$$

By rewriting the previous equation, we get

$$y(n) = \frac{1}{N} \sum_{m=0}^{N-1} x(m) \sum_{p=0}^{N-1} h(p) \sum_{k=0}^{N-1} e^{j2\pi k(n-m-p)/N}$$

But with q as an integer,

$$\sum_{k=0}^{N-1} e^{j2\pi k(n-m-p)/N} = \begin{cases} N & n-m-p = qN \\ 0 & \text{otherwise} \end{cases}$$

Thus, we can write

$$\sum_{k=0}^{N-1} e^{j2\pi k(n-m-p)/N} = N\delta(n-m-p=qN)$$

where $N\delta(n)$ is the impulse signal with a strength of N at $n=0$ and a strength of zero for $n \neq 0$. With the relation $\delta(n-m)f(n) = f(m)$, we can simplify the equation for $y(n)$ to get

$$y(n) = \sum_{m=0}^{N-1} x(m)\tilde{h}(n-m)$$

where $\tilde{h}(n)$ is a periodic repetition of $h(n)$. This equation is similar to the linear convolution equation when it is not limited to one period N. Based on what has been presented we will define circular convolution as

$$y(n) = \sum_{m=0}^{N-1} x(m)h(n-m) \tag{5.15}$$

and use the notation

$$y(n) = x(n) \otimes h(n)$$

to indicate circular convolution. With this result we can say now that circular convolution in the discrete domain is equal to multiplication in the DFT domain. That is,

$$y(n) = x(n) \otimes h(n) \leftrightarrow Y(k) = X(k)H(k) \tag{5.16}$$

Example 5.4

Consider the signals $x(n)$ and $h(n)$, each of length $N = 4$.

SOLUTION

To calculate the output $y(n)$ using circular convolution, we will write first the signals as

$$x(n) = \left\{ \begin{array}{cccc} x(0) & x(1) & x(2) & x(3) \end{array} \right\}$$

$$h(n) = \left\{ \begin{array}{cccc} h(0) & h(3) & h(2) & h(1) \end{array} \right\}$$

The output $y(n)$ will be of length 4 as well, and

$$y(0) = x(0)h(0) + x(1)h(3) + x(2)h(2) + x(3)h(1)$$

To find $y(1)$ we will move (shift) the last value in $h(n)$ and append it to the beginning of the array $h(n)$. The two arrays are now

$$x(n) = \left\{ \begin{array}{cccc} x(0) & x(1) & x(2) & x(3) \end{array} \right\}$$

$$h(n) = \left\{ \begin{array}{cccc} h(1) & h(0) & h(3) & h(2) \end{array} \right\}$$

and

$$y(1) = x(0)h(1) + x(1)h(0) + x(2)h(3) + x(3)h(2)$$

To find $y(2)$, we will list the two arrays by again performing the same shifting to get

$$x(n) = \left\{ \begin{array}{cccc} x(0) & x(1) & x(2) & x(3) \end{array} \right\}$$

$$h(n) = \left\{ \begin{array}{cccc} h(2) & h(1) & h(0) & h(3) \end{array} \right\}$$

with

$$y(2) = x(0)h(2) + x(1)h(1) + x(2)h(0) + x(3)h(3)$$

For the final value, $y(3)$, we will have

$$x(n) = \left\{ \begin{array}{cccc} x(0) & x(1) & x(2) & x(3) \end{array} \right\}$$

$$h(n) = \left\{ \begin{array}{cccc} h(3) & h(2) & h(1) & h(0) \end{array} \right\}$$

and

$$y(3) = x(0)h(3) + x(1)h(2) + x(2)h(1) + x(3)h(0)$$

Example 5.5

Consider the signals $x(n)$ and $h(n)$ with $N = 2$ and $x(0) = h(0) = 1$ and $x(1) = h(1) = 0$. Find the output $y(n)$ of the system given by this $h(n)$.

SOLUTION

Using circular convolution, we need to find $y(0)$ and $y(1)$. For $y(0)$, we have

$$x(n) = \left\{ \begin{array}{cc} 1 & 0 \end{array} \right\}$$

$$h(n) = \left\{ \begin{array}{cc} 1 & 0 \end{array} \right\}$$

$$y(0) = (1)(1) + 0(0) = 1$$

For $y(1)$, we have

$$x(n) = \left\{ \begin{array}{cc} 1 & 0 \end{array} \right\}$$

$$h(n) = \left\{ \begin{array}{cc} 0 & 1 \end{array} \right\}$$

$$y(1) = (1)(0) + 0(1) = 0$$

But we also know that $y(n)$ is the inverse DFT of $Y(k) = X(k)H(k)$. To find $X(k)$ and $H(k)$, we need to find $X(0)$, $X(1)$, $H(0)$, and $H(1)$.

$$X(0) = \sum_{n=0}^{1} x(n) e^{-\frac{j2\pi(0)}{N}} = x(0)(1) + x(1)(1) = 1 + 0 = 1$$

$$X(1) = \sum_{n=0}^{1} x(n) e^{-j2\pi(1)/N} = x(0)(1) + x(1)(-1) = 1$$

In the same way, we have $H(0) = 1$ and $H(1) = 1$. $Y(k)$ is the term-by-term multiplication of $X(k)$ and $H(k)$ and is

$$Y(0) = X(0)H(0) = 1(1) = 1$$

$$Y(1) = X(1)H(1) = 0(0) = 0$$

$y(n)$ is the inverse DFT of $Y(k)$ and is given by

$$y(n) = \frac{1}{N} \sum_{k=0}^{N-1} Y(k) e^{j2\pi nk/N}$$

with

$$y(0) = \frac{1}{2}\Big[Y(0)(1) + Y(1)(1)\Big] = \frac{1}{2}[1+1] = 1$$

$$y(1) = \frac{1}{2}\Big[Y(0)(1) + Y(1)(-1)\Big] = \frac{1}{2}[1-1] = 0$$

5.7.2   LINEAR CONVOLUTION

Consider the two signals, $x(n)$ and $h(n)$, each of length N_1 and N_2, respectively. If we append zeros to the end of $x(n)$ and $h(n)$ to make the length of both equal to $N_1 + N_2 - 1$, then the linear convolution is the same as the circular convolution. It means that we can use the DFT to find linear convolution if both signals have the length $N_1 + N_2 - 1$.

Example 5.6

Consider $x(n)$, where $x(0) = 1$ and $x(1) = 0$, and $h(n)$, where $h(0) = 1$ and $h(1) = 0$. Find $y(n) = x(n) * h(n)$ and $y(n) = x(n) \otimes h(n)$.

SOLUTION

First, we will make $x(n)$ and $h(n)$ each of length $N_1 + N_2 - 1 = 2 + 2 - 1 = 3$ by appending zeros at the end and write

$$x(n) = \Big\{ 1 \quad 0 \quad 0 \Big\}$$

$$h(n) = \Big\{ 1 \quad 0 \quad 0 \Big\}$$

With circular convolution, we have

$$y(0) = 1(1) + 0(0) + 0(0) = 1$$

To find $y(1)$, we list the arrays as

$$x(n) = \Big\{ 1 \quad 0 \quad 0 \Big\}$$

$$h(n) = \Big\{ 0 \quad 1 \quad 0 \Big\}$$

with $y(1) = 1(0) + 0(1) + 0(0) = 0$
  To find $y(2)$, we list the arrays again as

$$x(n) = \Big\{ 1 \quad 0 \quad 0 \Big\}$$

$$h(n) = \Big\{ 0 \quad 0 \quad 1 \Big\}$$

with $y(2) = 1(0) + 0(0) + 0(1) = 0$

So $y(n)$ with circular convolution is

$$y(n) = \left\{\begin{array}{ccc} 1 & 0 & 0 \end{array}\right\}$$

Using linear convolution, we get the same result.

5.7.3 Approximation to the Continuous Fourier Transform

We have seen in Section 5.4.3 that the approximation to the Fourier transform of the signal $x(t)$ is related to the DFT $X(k)$ as in the relation

$$X_c\left(\frac{2\pi}{NT_s}k\right) = \frac{1 - e^{-j2\pi k/N}}{j2\pi k/NT_s}X(k)$$

We can implement this approximation in MATLAB. Let us look at an example.

Example 5.7

Consider the signal

$$x(t) = \left\{\begin{array}{ll} 1 & 0 < t < 2 \\ 0 & \text{otherwise} \end{array}\right.$$

Plot the actual $X(w)$ and its approximation using the DFT.

SOLUTION

The actual Fourier transform of $x(t)$ is

$$X(w) = 2\text{sinc}(w)e^{-jw} = 2\frac{\sin(w)}{w}e^{-jw}$$

First we need to sample $x(t)$ to get $X(k)$ and then we need to scale $X(k)$ to get the approximation to $X(w)$, the actual continuous Fourier transform. In the MATLAB script Example5_7, we will use a sampling period of 0.1 s and a number of samples $N = 2^7 = 128$. Note that to use the fft function with MATLAB, N must be 2 raised to an integer power.

The plots are shown in Figure 5.3. If we increase the number of samples N to $2^8 = 256$, we have the better approximation as seen in Figure 5.4.

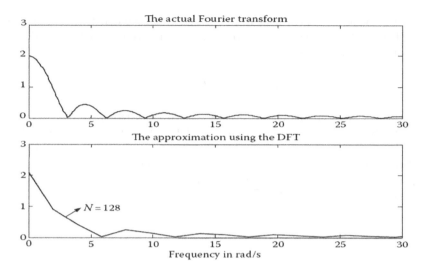

FIGURE 5.3 Plots for Example 5.7.

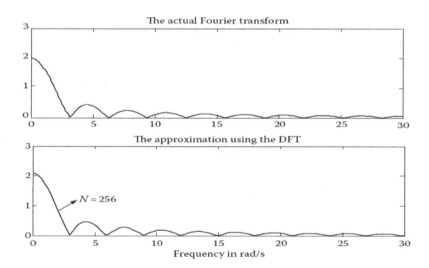

FIGURE 5.4 Plots for Example 5.7.

5.7.4 APPROXIMATION TO THE COEFFICIENTS OF THE FOURIER SERIES AND THE AVERAGE POWER OF THE PERIODIC SIGNAL $x(t)$

Given a periodic continuous signal, $x(t)$, with the period T, the Fourier series approximation is calculated first by evaluating the Fourier series coefficients

$$c(k) = \frac{1}{T} \int_T x(t) e^{-jw_0 kt} dt \qquad (5.17)$$

For the computer to approximate this integration, we need to represent $x(t)$ in a discrete form. Let us sample $x(t)$ at $t = nT_s$, where T_s is the sampling interval and n is an integer. Thus, we will have

$$x(n) = x(t)\big|_{t=nT_s}$$

Remember that we are integrating over the period T in continuous time. In discrete time, we need to subdivide this period T into intervals of width T_s and we will have N of these intervals. Thus, we write $T = NT_s$. If we make T_s very small (N very large) and approximate the area under $x(t)e^{-jw_0kt}$ in these intervals, we can have a good approximation to the integral equation given.

With this process, w_0 will be

$$w_0 = \frac{2\pi}{T} = \frac{2\pi}{NT_s}$$

and $c(k)$ will be

$$c(k) = \frac{1}{NT_s}\left[\sum_{n=0}^{N-1} x(nT_s)e^{-j(2\pi/NT_s)knT_s}\right]T_s$$

where the integration becomes summation and hence dt becomes T_s. Finally, the equation for $c(k)$ becomes

$$c(k) = \frac{1}{N}\sum_{n=0}^{N-1} x(nT)e^{-j2\pi kn/N} \tag{5.18}$$

Equation (5.18) is $1/N$ times the DFT of $x(nT_s)$. Thus,

$$X(k) = Nc(k)$$

where $X(k)$ is the DFT of $x(nT_s)$. Remember also that

$$c(0) = \frac{1}{T}\int_0^T x(t)\,dt$$

is the average value of $x(t)$. Hence $X(0)/N$ is an approximation to the average value of $x(t)$. The average power in the periodic signal $x(t)$ is given as

$$P_a = \frac{1}{T}\int_0^T |x(t)|^2\,dt \tag{5.19}$$

and its approximation using the Fourier series coefficients is

$$P_a \approx \sum_{k=-\infty}^{\infty} |c(k)|^2 \tag{5.20}$$

But with $X(k) = Nc(k)$, we have the average power approximated as

$$P_a \approx \frac{1}{N^2} \sum_{k=0}^{N-1} |X(k)|^2 \tag{5.21}$$

Example 5.8

Consider the signal

$$x(t) = \sin(2\pi t)$$

Find the average value of the signal, its average power, and its Fourier series coefficients as well as their approximations.

SOLUTION

The signal is periodic with period $T = 1$ s. The average value is

$$c(0) = \frac{1}{1} \int_0^1 \sin(2\pi t)\,dt = -\left(\frac{1}{2\pi} - \frac{1}{2\pi}\right) = 0$$

The average power is given as

$$P_a = \frac{1}{1} \int_0^1 |\sin(2\pi t)|^2\,dt = \frac{1}{2}$$

The other Fourier series coefficients can be found by using MATLAB as in the script Example5_81.
   The result is

```
ck = 0
ck = -1/2 * i
ck = 0
ck = 0
```

You can see from this result, as expected, that the sine wave has only one frequency component, that $c(0) = 0$, the average value and the $|c(1)| = 1/2$ is the only frequency component that is nonzero.
   We can use the DFT to find an approximation to $c(0)$, the average power and the other frequency components. First we see that $x(t)$ is periodic with $T = 1$ s.

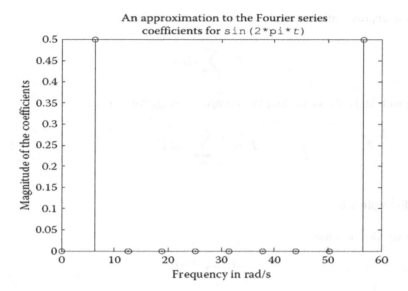

FIGURE 5.5 Plots for Example 5.8.

We will divide this T interval into equally spaced intervals. With the frequency of the signal $x(t)$ equal to 1 Hz, we choose f_s to be 10 Hz. Thus, with $f_s = 10$ Hz, $T_s = .1$ s. The number of samples we will take from $x(t)$ in the T interval is

$$N = \frac{T}{T_s} = \frac{1}{0.1} = 10 \text{ samples}$$

The MATLAB script Example5_82 will be used.
    The average value as expected is zero and the average power is 1/2. The plot is shown in Figure 5.5.

5.7.5 Total Energy in the Signal $x(n)$ and $x(t)$

The total energy in the signal $x(n)$ for $n = 0, 1, ..., N - 1$ is given as

$$E_t = \sum_{n=0}^{N-1} |x(n)|^2 = \frac{1}{N} \sum_{k=0}^{N-1} |X(k)|^2 \qquad (5.22)$$

If we are given a nonperiodic signal $x(t)$, we can talk about its total energy. If we sample $x(t)$ at the sampling rate f_s satisfying the Nyquist sampling rate, then the approximate energy in the signal is

$$E_t = \sum_{n=n_1}^{n_2} |x(nT_s)|^2 T_s = \frac{T_s}{N} \sum_{k=0}^{N-1} |X(k)|^2$$

n_1 and n_2 are the integer values for which $x(t)$ is approximately zero for $t < n_1$ and $t > n_2$.

Example 5.9

Consider the signal $x(t) = e^{-t}$ for $t > 0$. What is the total energy in the signal?

SOLUTION

The total energy in the signal is

$$E_t = \int_0^\infty \left|\left(e^{-t}\right)\right|^2 dt = \frac{1}{2}$$

We can compute this value of 1/2 using the DFT. But first let us look at the continuous Fourier transform of $x(t)$

$$X(w) = \frac{1}{1 + jw}$$

The magnitude of $X(w)$ is given as

$$\left|X(w)\right| = \frac{1}{\sqrt{1 + w^2}}$$

We can see that for $w > 2\pi(50)$, the magnitude of $X(w)$ approaches zero. Thus let $f_s = 2(50) = 100$ Hz be the sampling frequency. We can also see that for $t > 5$, the signal $x(t)$ approaches zero as well. Therefore, we will choose the time interval of 5 s and sample $x(t)$ in this interval. In this case, the number of samples N is $T/T_s = 5(100) = 500$.

We can use MATLAB to calculate the total energy in $x(t)$ as in the script Example5_91.
  to get

Etotal = 0.505 using the time domain
Effttotal = 0.506 using the DFT

Note that if we were given a discrete signal $x(n)$ such that

$$x(0) = 1 \quad x(1) = 0 \quad x(2) = 1 \quad x(3) = 1 \quad x(4) = 0$$

then

$$E_t = \sum_{n=0}^{N-1} |x(n)|^2 = \left(x(0)\right)^2 + \left(x(1)\right)^2 + \left(x(2)\right)^2 + \left(x(3)\right)^2 + \left(x(4)\right)^2 = 1 + 0 + 1 + 1 + 0 = 3$$

MATLAB can be used also to compute the total energy in the signal as in the script Example5_92.
  The result will be

Etotal = 3
Effttotal = 3

5.7.6 Block Filtering

A digital filter can be described by the impulse response $h(n)$. $h(n)$ can change the magnitude of the input signal as well as produce a phase shift. In using the DFT to calculate $y(n)$, the output of the filter $h(n)$, notice that you need the entire $x(n)$ present. In case of huge $x(n)$, this will cause a considerable delay in obtaining $y(n)$. If we break $x(n)$ into blocks of data, we can use the DFT to produce blocks of the output $y(n)$. These blocks then can be arranged to produce the total output $y(n)$. This method will produce outputs faster and will reduce the waiting time for incoming $x(n)$ samples. An example of how to perform this method will be given in the EOCE section later.

5.7.7 Correlation

Correlation is often used in the detection of a target in a radar signal. It can also be used in the estimation of the frequency content of a certain signal. Cross-correlation is the correlation between two different signals and auto-correlation is the correlation with the signal itself. The cross-correlation is given by the relation

$$R_{x1x2}(p) = \sum_{n=-\infty}^{+\infty} x_1(n) x_2(p+n) \tag{5.23}$$

Equation (5.23) is similar to the convolution equation with the only difference being that $x_2(n)$ is shifted but not reflected.

In dealing with discrete signals, we use N samples and assume that the signals are periodic with period NT_s. With periodic signals and if $0 \le n \le N - 1$, we have

$$x(-n) = x(-n+N)$$

The cross-correlation equation can be written then as

$$R_{x1x2}(p) = \sum_{n=-\infty}^{+\infty} x_1(n) x_2(p-(-n)) = x_1(-n) * x_2(n) \tag{5.24}$$

If we take the DFT of Equation (5.24), we will have

$$R_{x1x2} \leftrightarrow X_1(-k) X_2(k) \tag{5.25}$$

and then the inverse DFT of $X_1(-k) X_2(k)$ is $R_{x1x2}(p)$. If $x_2(n)$ is real, then

$$X_1(-k) = X_1 *(k)$$

Finally, we have

$$R_{x1x2}(p) \leftrightarrow X_1(k) * X_2(k) \tag{5.26}$$

Note that $R_{x1x2}(p) \neq R_{x2x1}(p)$. It can be shown that

$$R_{x1x2}(p) \leftrightarrow x_1(k)x_2(k)^*$$ (5.27)

One important application of auto-correlation is the estimation of the energy spectrum density of the signal $x(n)$. The auto-correlation process helps to eliminate the noise if it is present in a certain signal. The energy spectrum density estimate is

$$R_{xx}(p) = \frac{1}{N}\sum_{n=0}^{N-1} x(n)x(p+n) \; 0 \leq p \leq N-1$$ (5.28)

From this relation, we have

$$R_{xx}(p) \leftrightarrow \frac{1}{N}X(k)X(k)^*$$ (5.29)

We will give an example on how to use this relation later in the EOCE.

5.8 SOME INSIGHTS

5.8.1 DFT IS THE SAME AS THE fft

The DFT is not an approximation to the fft; it is the same as the fft. The fft is a fast and an efficient way of calculating the DFT.

5.8.2 DFT POINTS ARE THE SAMPLES OF THE FOURIER TRANSFORM OF $x(n)$

The Fourier transform of the discrete signal $x(n)$ for $n = 0, 1, ..., N-1$ is

$$X(e^{j\theta}) = \sum_{n=0}^{N-1} x(n)e^{-j\theta n}$$

If we sample $X(e^{j\theta})$ on the unit circle with $\theta = 2\pi k/N$ for $k = 0, 1, ..., N-1$, then

$$X(k) = X(e^{j\theta})\big|_{\theta=2\pi k/N}$$

5.8.3 HOW CAN WE BE CERTAIN THAT MOST OF THE FREQUENCY CONTENTS OF $x(t)$ ARE IN THE DFT?

To get $X(k)$ from $x(t)$, we need first to sample $x(t)$ making sure that f_s, the sampling frequency used to produce $x(n)$, is at least twice f_m, the maximum frequency in $x(t)$.

This is necessary to avoid aliasing. Next, N samples should be collected from $x(t)$ in NT_s s, where $T_s = 2\pi f_s$. The number of samples N is inversely related to the frequency resolution Δf according to

$$\Delta f = \frac{2\pi}{N}$$

The smaller the frequency resolution is, the better is the process of detecting most of the frequency components in $x(t)$.

5.8.4 IS THE CIRCULAR CONVOLUTION THE SAME AS THE LINEAR CONVOLUTION?

The circular convolution between the two signals $x_1(n)$ and $x_2(n)$ will be the same as the linear convolution if we append zeros to the end of $x_1(n)$ and $x_2(n)$ so that the number of samples in $x_1(n)$ and $x_2(n)$ will be $N_1 + N_2 - 1$, where N_1 is the number of samples in $x_1(n)$ and N_2 is the number of samples in $x_2(n)$.

5.8.5 IS $|X(w)| \approx |X(k)|$?

Due to the inherent factor $1/T_s$ in calculating the Fourier transform of discrete signals, the magnitude of the DFT, $|X(k)|$ should be multiplied by T_s to get the approximation to $|X(w)|$. Note that this approximation is not as good as the approximation we discussed earlier in this chapter.

5.8.6 FREQUENCY LEAKAGE AND THE DFT

In the development of the DFT, the signal $x(nT_s)$ is made periodic first and then multiplied by the rectangular window of unity magnitude that extends from $n = 0$ to $n = N - 1$. The result of this multiplication is the signal $x(n)$. To understand this better, consider the signal $x(t) = \sin(t)$. This signal is periodic and there is no need to make it periodic. Its Fourier transform consists of one component at $w = 1$. If we truncate $x(t)$ by multiplying it with a rectangular window that is centered at $t = 0$, the Fourier transform of $\sin(t)$ multiplied by this window will result in a sinc-shape graph centered at $w = 1$. From this you can see that the frequency content of $x(t)$ after this multiplication is distorted. This is referred to as frequency leakage.

To have better approximation, different windows can be used. Two well-known windows are the Hanning and the Hamming windows. These windows do not have sharp cuts at the edges; they gradually approach zero. The two windows are implemented in MATLAB.

5.9 END-OF-CHAPTER EXAMPLES

EOCE 5.1

Consider the discrete time system represented by

$$h(n) = (0.5)^n \quad n = 0, 1, 2, \cdots, 10$$

and consider an input to the system given by

$$x(n) = u(n) \quad n = 0, 1, 2, \cdots, 5$$

Find the output $y(n)$ using the DFT method.

SOLUTION

We have $N_h = 11$ samples for $h(n)$ and $N_x = 6$ samples for $x(n)$. So we need to add $6 - 1 = 5$ zeros to $h(n)$ and $11 - 1 = 10$ zeros to $x(n)$ so that each signal will have $(11 + 6) - 1 = 16$ samples. At this point we can use the DFT to calculate $y(n)$. The MATLAB script EOCE5_1 will compute $y(n)$ using the function conv and it will also compute $y(n)$ by inverse transforming $Y(k)$ for comparison.

The result is plotted in Figure 5.6. Note that we have $h(n)$ and $x(n)$ each of length 16. This means that the radix-2 fft was used. If we want the faster and more efficient radix-2 to be used, each signal should have 2^p as its length where p is a positive integer.

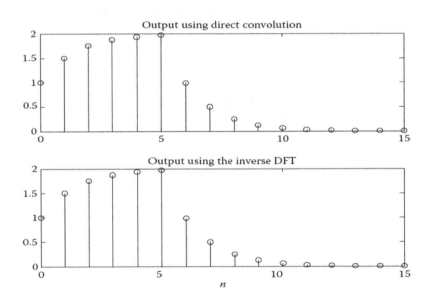

FIGURE 5.6 Plots for EOCE 5.1.

EOCE 5.2

Find an approximation to the magnitude of the Fourier transform of

$$x(t) = e^{-10t}u(t)$$

SOLUTION

To use the DFT to approximate the Fourier transform of $x(t)$, we need to sample $x(t)$ first. Let us look at the Fourier transform of $e^{-10t}u(t)$ so that we can find the frequency at which $X(w)$ is approaching zero. The Fourier transform is

$$X(w) = \frac{1}{jw + 10}$$

and the magnitude of $X(w)$ is

$$|X(w)| = \frac{1}{\sqrt{w^2 + 100}}$$

For $w = 30\pi$ rad/s, the magnitude of $X(w)$ is approximately zero. Thus we choose $w_s \geq 2w = 60\pi$. Let us choose $w_s = 300\pi$ with $f_s = 150$ Hz and $T_s = 1/150$s. Note that at $t = 1$ s, $x(t)$ is approximately zero. Let us choose $NT_s = 1$ or $N = 1/T_s = 150$ samples. We have established a lower limit on N of 150. To use the radix-2 fft, N must be an integer power of 2. The next N then will be $2^8 = 256$.

We have seen in this chapter two ways to perform this approximation. In the first method, the approximation was given by

$$\frac{1 - e^{-j2\pi k/N}}{j(2\pi k/NT_s)} X(k)$$

and in the second method, the approximation to $|X(w)|$ was $T_s (X(k))$. In the script EOCE5_2, we use the two methods.

The plots are shown in Figure 5.7.

EOCE 5.3

Consider the signal

$$x(t) = \sin(2\pi 2000t) + \frac{1}{2}\sin(2\pi 4000t)$$

1. Use the DFT to approximate $|X(w)|$ with $N = 16$, and choose f_s such that the Nyquist criterion is observed and $f_s/N = f_1/k_1 = f_2/k_2$, where k_1 and k_2 are integers and f_1 and f_2 are the frequencies of the two components in $x(t)$, 2000 and 4000 Hz.
2. Repeat while $f_s/N \neq m_1 f_1 \neq m_2 f_2$.
3. Can you do a correction to the approximation?

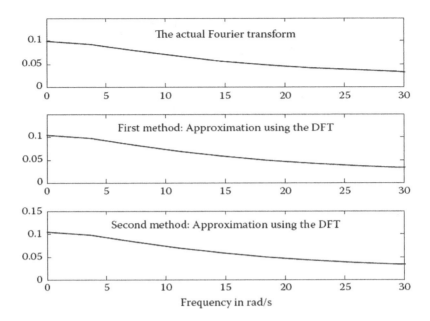

FIGURE 5.7 Plots for EOCE 5.2.

SOLUTION

1. The highest frequency in $x(t)$ is $f_2 = 4000$ Hz. With $f_s = 16,000$ Hz,

$$\frac{f_s}{N} = \frac{16,000}{16} = 1,000 = \frac{f_1}{2} = \frac{f_2}{4}$$

Therefore, on the frequency axis of the magnitude of $|X(k)|$, we will see that the frequency content of the first component at $f_1 = 2000$ Hz corresponds to $k_1 = 2$, and the frequency content of the second component at $f_2 = 4000$ Hz corresponds to $k_2 = 4$. In this case, the Fourier transform of $x(t)$, $X(w)$, will match exactly the DFT of $x(n)$, $X(k)$. The MATLAB script to prove that is EOCE5_31.

The plot is shown in Figure 5.8. The magnitude of $X(2) = 8$ is the magnitude of the 2000 Hz term times N divided by 2. This is true because the first term in $x(t)$ has no dc components.

2. If we sample at $f_s = 10,000$ Hz (satisfying the Nyquist rate) and with $N = 16$ we have

$$\frac{f_s}{N} = \frac{10,000}{16} \neq \frac{f_1}{k_1} \neq \frac{f_2}{k_2}$$

for any k_1 or k_2. In this case there will be distortion in the frequency spectrum using the DFT. This is to say that the 2000 and 4000 Hz frequencies will not be observed at any value of k in the DFT magnitude plot. For $f_s = 10,000$ Hz, the approximation to $x(t)$ will be as seen in Figure 5.9.

3. If we use the Hanning window instead of the rectangular window inherent in the DFT development, we will get some improvements as seen in Figure 5.9. The MATLAB script for using the Hanning and the rectangular windows is EOCE5_32.

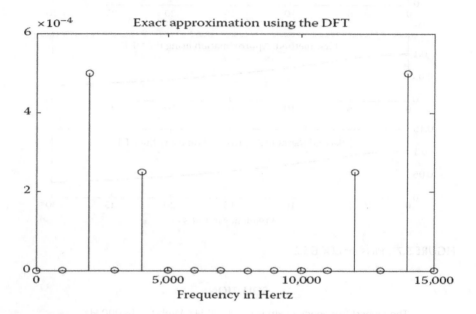

FIGURE 5.8   Plot for EOCE 5.3.

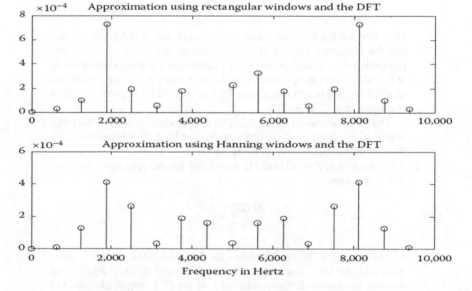

FIGURE 5.9   Plots for EOCE 5.3.

EOCE 5.4

Consider the signals $x(t)$ and $h(t)$ that are the input and the impulse response of a linear system

$$h(t) = e^{-100t}$$

$$x(t) = \sin\left(2\pi(1000)t\right)$$

Find the output $y(n)$ and display the frequency spectrum of the input and the output using the DFT.

SOLUTION

The input $x(t)$ is periodic with period $T = 0.001$. We will sample $x(t)$ at $f_s = 8000$ Hz satisfying the Nyquist rate. For $f_s = 8000$ Hz we have $T_s = 1/8000$ s. We will sample $x(t)$ for $T = 0.001$ s. To choose a value for N we use the relation $NT_s = T$ or $N = T/T_s = 0.001/1/8000 = 8$ samples.

For $h(t) = e^{-100t}$ we see that if $t = 0.1$ s, $h(t)$ will be very close to zero in value. The Fourier transform of $h(t)$ is $H(w)$ and is

$$H(w) = \frac{1}{jw + 100}$$

with a magnitude of

$$|H(w)| = \frac{1}{\sqrt{w^2 + (100)^2}}$$

With $w = 2\pi(20)$ rad/s, $|H(w)|$ will approach zero. So we choose w_s as $w_s > 2(2\pi)$ (20) to satisfy the Nyquist rate. Let us make $w_s > 2\pi(200)$. Thus $f_s = 200$ and $T_s = 1/200$ s. So for a time length of 0.1 s (at which $h(t) \approx 0$) and with $T_s = 1/200$, we have

$$N\left(\frac{1}{200}\right) = 0.1 \text{ or } N = 0.1(200) = 20 \text{ samples}$$

Now we will find the DFT for $x(t)$ with $N = 8$ and the DFT for $h(t)$ with $N = 20$. To find $y(n)$ we will use convolution by making $x(n)$ and $h(n)$ both of length $n_1 + n_2 - 1$, where n_1 and n_2 are the number of samples in $x(n)$ and $h(n)$, respectively. To do that we will pad $x(n)$ and $h(n)$ by zeros. The script EOCE5_4 will do that.

The plots are shown in Figure 5.10. Notice in this MATLAB script that the DFT of $x(n)$, $h(n)$, and the inverse transform of $Y(k)$ were calculated using the relation

```
N = length (xn) + length (hn) -1;
```

But 27 is not an integer power of 2 and hence the radix-2 fft was not used. To use the radix-2 fft, we need not pad $x(n)$ or $h(n)$ with zeros since padding will be made by calling the fft function

```
Xk = fft (xn, N);
```

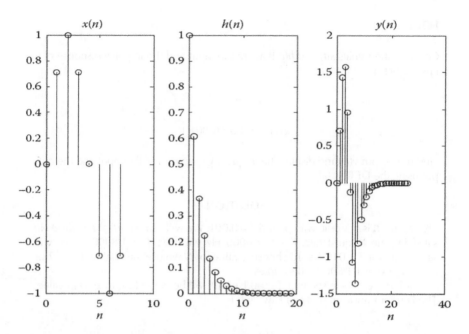

FIGURE 5.10 Plots for EOCE 5.4.

In this case $x(n)$ will be made of length N where $(N - \text{length}(xn))$ is the number of the padded zeros. We can use this method with $h(n)$ as well. When we use the command

```
Yk = Xk * Hk';
```

Yk will be of length N as well.

EOCE 5.5

Consider the discrete system where the input $x(n)$ is the pulse defined as

$$x(n) = \begin{cases} 1 & 0 \leq n \leq 5 \\ 0 & \text{otherwise} \end{cases}$$

and the impulse response is defined as

$$h(n) = (.2)^n u(n)$$

Find the output $y(n)$ using the DFT.

SOLUTION

The signals here are already sampled. We can see that $x(n)$ is zero for $n \geq 6$. Thus the number of samples in $x(n)$ is 6. For $h(n)$, there is no real integer that makes $h(n)$

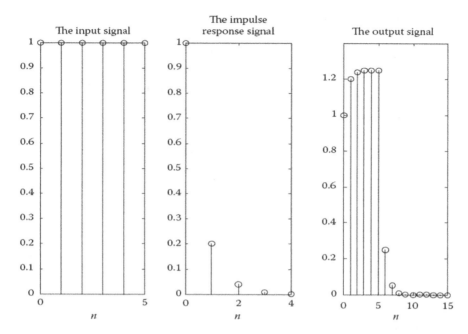

FIGURE 5.11 Plots for EOCE 5.5.

zero. But if $n = 5$, $h(n)$ is close to zero. Thus $h(n)$ will have a value close to zero for $n \geq 5$. For this $h(n)$ we will have five samples. To use radix-2 fft, $N_1 + N_2 - 1$ should be an integer multiple of 2. $N_1 + N_2 - 1 = 6 + 5 - 1 = 10 \neq 2^p$ for any integer p. The next integer 2^p greater than 10 is $16 = 2^4$. So we need to use at least 16-point DFT to find $y(n)$ using the radix-2 fft. The MATLAB script EOCE5_5 will be used.

The plot is shown in Figure 5.11.

EOCE 5.6

Suppose that we have 2 s of the signal $x(t)$. Let us sample $x(t)$ at $f_s = 1000$ Hz.

1. Find the maximum frequency that can be present if there is to be no aliasing.
2. What analogue frequencies are present in the DFT?
3. What is the frequency resolution?

SOLUTION

1. With $f_s = 1000$ Hz, from the Nyquist condition, we have

$$f_s \geq 2f_m \text{ or } f_m = \frac{f_s}{2} = \frac{1000}{2} = 500 \text{ Hz}$$

2. The frequency spacing (the frequency resolution) is given by

$$\frac{f_s}{N} = \frac{1}{T_r} = \frac{1}{2} = 0.5\,\text{Hz}$$

where T_r is the total time of observing the signal and is called the record length.

3. From part 2 with $f_s/N = 0.5$, we get

$$N = \frac{f_s}{.5} = \frac{1000}{.5} = 2000$$

The frequencies in the DFT are at

$$\frac{mf_s}{N} = m\frac{1000}{2000} = \frac{1}{2}m\,\text{Hz}$$

where m is an integer. So they are at

$$0, \frac{1}{2}, 1, \frac{3}{2}, 2, \cdots, 500, -499, -\frac{1}{2}$$

EOCE 5.7

Consider the two signals $x_1(n) = x_2(n)$ with

$$x_1(0) = 1 \; x_1(1) = 1 \; x_1(2) = 0 \text{ and } x_1(3) = 1$$

Find the cross-correlation between $x_1(n)$ and $x_2(n)$, R_{x1x2} and R_{x2x1}.

SOLUTION

Note that $x_1(n)$ has four samples as well as $x_2(n)$. To use the DFT to find the cross-correlation, we need to make both $x_1(n)$ and $x_2(n)$ of length $4 + 4 - 1 = 7$ samples. We will pad $x_1(n)$ with three zeros and we will do the same for $x_2(n)$.

The MATLAB script EOCE5_71 will accomplish this.

The plots are shown in Figure 5.12. We could have used the MATLAB function conv to find the same result. The script EOCE5_72 can be used.

The plots are shown in Figure 5.13.

EOCE 5.8

Use MATLAB to generate an N-points random signal $x_1(n)$, then add to it the N-points samples of the 1000 Hz signal $x_2(n)$ where

$$x_2(n) = \sin\left(\frac{2\pi(1,000)n}{10,000}\right)$$

Find the energy spectral estimate of $x(n)$ where

$$x(n) = x_1(n) + x_2(x)$$

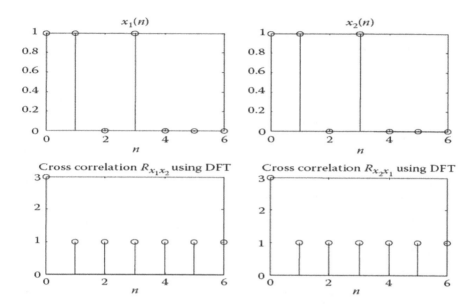

FIGURE 5.12 Plots for EOCE 5.7.

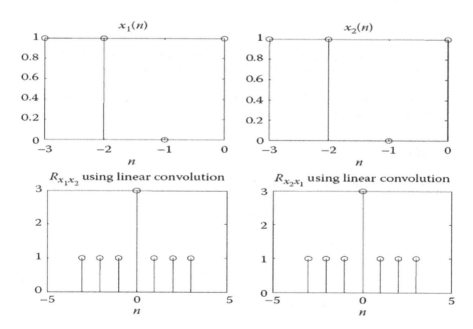

FIGURE 5.13 Plots for EOCE 5.7.

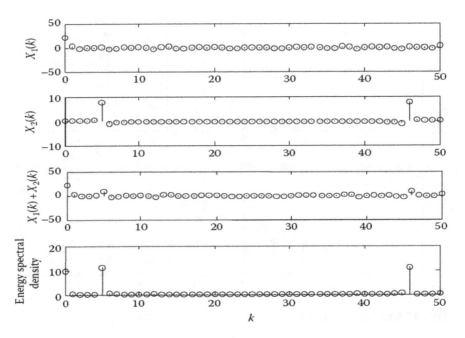

FIGURE 5.14   Plots for EOCE 5.8.

SOLUTION

We will use MATLAB to generate 51 random values that are uniformly distributed between zero and one. The MATLAB script to calculate the spectral energy estimate is EOCE5_8.

The plots are shown in Figure 5.14. You can clearly see that the 1 kHz signal is standing out while the noise is attenuated.

EOCE 5.9

Consider a discrete system with input

$$x(n) = e^{-n} \quad 0 \le n \le 20$$

and impulse response

$$h(n) = 1 \quad 0 \le n \le 1$$

Find the output $y(n)$ using block filtering.

SOLUTION

The impulse response has 2 values and the input has 21 values. If we want to use 8-point DFT then we can divide $x(n)$ into 3 blocks each having 7 values in it. Thus $7 + 2 - 1 = 8$ is the length of the DFT output as the result of convolving the

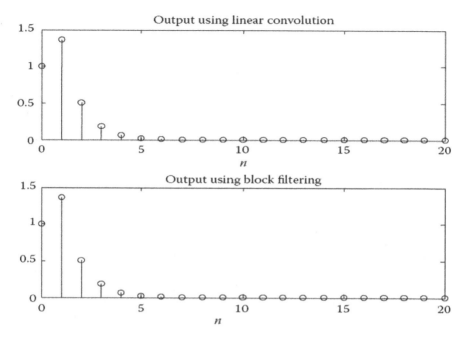

FIGURE 5.15 Plots for EOCE 5.9.

7-values input with the 2-values impulse response. Let us call the input blocks $x_1(n)$, $x_2(n)$, and $x_3(n)$. The output blocks are then $y_1(n)$, $y_2(n)$, and $y_3(n)$. Note that for 7-values input we will have 8-values output. To find out the final output, we will form the array $y(n)$ as a series of the subarrays $y_1(n)$, $y_2(n)$, and $y_3(n)$, where the eighth value of $y_3(n)$ will be added to the first value of $y_2(n)$ and the eighth value of $y_2(n)$ will be added to the first value of $y_3(n)$. This array of $y(n)$ can be viewed as

$$y(n) = \left[y_1(0) \cdots y_1(6) \big(y_1(7) + y_2(0) \big) y_2(1) \cdots y_2(6) \big(y_2(7) + y_3(0) \big) y_3(1) \cdots y_3(7) \right]$$

The MATLAB script EOCE5_9 will produce $y(n)$ using block filtering with the DFT and regular linear convolution.

The plots are shown in Figure 5.15.

EOCE 5.10

High resolution and *dense spectrum* are two misleading terms. Next we consider an example to show you the difference. Consider the signal

$$x(n) = \sin(.37\pi n) + \sin(.55\pi n)$$

It is clear that $x(n)$ has the two frequencies at $.37\pi$ and $.55\pi$. Let us take only 20 samples of the 100 samples from $x(n)$ and find the spectrum. Then let us pad the 20-sample signal with 80 zeros and find the spectrum again. Next let us take all 100 samples from $x(n)$ and find the spectrum.

SOLUTION

We will use the MATLAB script EOCE5_10 to accomplish this task.

The plots are shown in Figure 5.16. Notice that with only 20 samples of $x(n)$ and no zero padding, the spectrum is not dense as it is in the case when 20 samples are taken with 80 zeros padded. Padding with zeros will make the spectrum denser. This does not mean that the DFT with zero padding will tell more about the frequency contents of the signal. As it is seen in the plot, with few samples and no zero padding and with few samples and zero padding, the magnitude plots are distorted and do not have values of the exact two frequencies of .37π and .55π. However, by taking more points from $x(n)$, we can see the spectrum emphasized at the .37π and .55π frequencies.

Thus we can say that zero padding can only make spectrum denser which does not mean good frequency resolution. More points or samples will produce more frequency components and yet good frequency resolution. This is not to say that good frequency resolution comes from more samples only. You need to consider a good time span over which the signal is well known. If the signal is periodic, you can sample over one period. Taking more samples within this period gives good frequency resolution and yet more frequencies will be detected using the DFT. If the signal is not periodic but its amplitude approaches zero as time reaches a certain limit, then you need to sample the signal for the period of time up to that limit. This is what is known as the record length in digital signal processing.

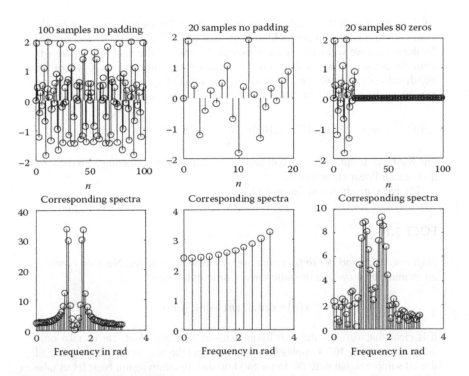

FIGURE 5.16   Plots for EOCE 5.10.

EOCE 5.11

In EOCE 1.11, we have acquired an audio signal using the data acquisition toolbox of MATLAB. Here we would like to see how we can FFT the audio signal to identify its frequency contents.

SOLUTION

Let us repeat EOCE 1.11 here and modify to get and plot the FFT.

To complete this task, you will need a PC microphone or your laptop built-in microphone and a sound card that is also found on your PC or laptop. The sound card will work as the analogue-to-digital converter as will be discussed later in the chapters.

Let us start by acquiring 3 s of audio using the sound card. What we speak is an analogue signal and so the sound card will convert the speech to digital samples. Let us use 8000 Hz as the sampling rate on the sound card.

We will first create an analogue input object for the sound card. At the MATLAB prompt write

```
AI = analoginput ('winsound');
```

to create an input communication channel with the sound card. Then you need to add an input channel to the input object AI. This is done with the MATLAB command

```
addchannel (AI, 1);
```

to add a single channel to the input object.

We need to specify a sampling rate (this topic will be revisited again in later chapters for complete discussion). We will use 8000 Hz for that. We will write

```
Fs=8000;
Set (AI, 'SampleRate',Fs);
```

to define the sample rate to MATLAB and associate that with the input object AI.

We need to speak through the mice for a limited time of 3 s and tell the channel how many samples per trigger we wish to collect. We do that with

```
Trig_duration = 3;
Set (AI, 'SamplesPerTrigger',Trig_duration*Fs);
```

At this stage we can start collecting data for the audio signal we will speak using the mice. Before you hit enter after you type the following command, start speaking for about 5 s.

```
start (AI);
```

To collect the data samples of the words you just spoke use the command

```
data = gatdata (AI);
```

Once you have the data, you can apply the FFT algorithm.

```
datafft = abs (fft (data));
```

and in dB, the magnitude is calculated as

```
magdatafft = 20*log10 (datafft);
magdatafft = mag (1:end/2);
```

We can plot the fft magnitude as in

```
plot (magdatafft);
```

We can also see what is the highest and lowest frequency in the spoken words we spoke. We write

```
max (magdatafft)
min (magdatafft)
```

It is always a good practice to delete the AI object after we finish. Write the command

```
delete (AI);
```

to delete the AI object

At this time, you can plot the data you just collected.

5.10   END-OF-CHAPTER PROBLEMS

EOCP 5.1

A continuous time signal has f_m as its highest frequency. If $f_m = 1$ kHz and we desire sampling the signal at 10 times f_m, what would be the record length and the number of samples if the frequency resolution is to be 10 Hz?

EOCP 5.2

Find the circular convolution between the signals

1. $u(n)$ for $0 \leq n \leq 5$ and $e^{-n/.1}$ for $0 \leq n \leq 5$
2. $5\sin(n\pi/3)$ for $0 \leq n \leq 4$ and e^{-n} for $0 \leq n \leq 4$
3. $e^{-n/2}$ and itself for $0 \leq n \leq 10$
4. $x(n) = 1$ and itself for $0 \leq n \leq 5$
5. $\cos(n\pi/6)$ for $0 \leq n \leq 6$ and $\delta(n)$ for $0 \leq n \leq 6$

EOCP 5.3

1. Find the DFT $X(e^{j\theta})$ of $x(n) = e^{-n/3}$ for $0 \leq n \leq 7$.
2. Sample $X(e^{j\theta})$ at $\theta = 2\pi k/N$ with $N = 16$.
3. Find the first four values $X(0), \ldots, X(3)$ using the DFT equation.
4. Find all eight values for $X(k)$ and compare with the values found in part 2.
5. If $x(n)$ in part 1 is an input to the linear system given by $h(n) = e^{-n/6}$ for $0 \leq n \leq 3$, use convolution to find $y(n)$, the output of the system.
6. Use the DFT to find $y(n)$ in part 5.

EOCP 5.4

Consider the signal

$$x(n) = \begin{cases} \cos(n\pi/6) & 0 \le n \le 5 \\ 0 & \text{otherwise} \end{cases}$$

1. Let $N = 8$, 16, 32, and 64. Find the spectrum for $x(n)$ using the DFT.
2. What conclusion can you draw by completing part 1?
3. If $x(n) = \cos(n\pi/6)$ for $0 \le n \le 63$, plot the spectrum for $x(n)$ in this case.
4. Compare the results of part 1 and part 3.

EOCP 5.5

Find the DFT of the following signals where n is taken in the interval $0 \le n \le N - 1$. A is a constant. Use 3 different appropriate N values.

1. $A\delta(n)$
2. A
3. $A \sin(2\pi n/N)$
4. $A \cos(2\pi n/N)$

EOCP 5.6

Consider the signal

$$x(t) = \sin(600\pi t) + \sin(1000\pi t)$$

1. What is the period of $x(t)$?
2. What is the minimum sampling frequency?
3. Sample $x(t)$ for one period at $f_s = 10 f_m$ and plot $x(n)$.
4. Find the DFT of $x(n)$ in 3.
5. Repeat part 3 over two periods. What do you observe? Keep N as in part 3.
6. Find the DFT of $x(n)$ in part 5. What do you notice?

EOCP 5.7

Use the DFT to find the energy spectrum density for the signals

1. $x(n) = \sin\left(\dfrac{2\pi n}{11}\right)$ $0 \le n \le 15$

2. $x(n) = e^{-n/3}\sin\left(\dfrac{2\pi n}{11}\right)$ $0 \le n \le 15$

EOCP 5.8

Use the DFT to find the cross-correlation between the signals given in EOCP 5.7.

EOCP 5.9

Use the DFT to approximate the Fourier transform for the signals

1. $x(t) = e^{-t}u(t)$
2. $x(t) = 20\cos\left(\dfrac{2\pi t}{13}\right)$
3. $x(t) = \begin{cases} 1 & 0 \le t \le 1 \\ 0 & \text{otherwise} \end{cases}$

EOCP 5.10

Consider the system

$$h(n) = e^{-n} \ 0 \le n \le 10$$

If $x(n) = u(n)$ for $0 \le n \le 5$, find $y(n)$ using linear convolution via the DFT.

EOCP 5.11

Consider the system

$$h(t) = e^{-t/3}u(t)$$

1. If the input is $x(t) = e^{-5t}u(t)$, use the DFT to find $y(n)$.
2. If the input is $x(t) = 10\sin(2\pi(500)t) + 10$, find the output $y(n)$ using the DFT.
3. Is there any dc component in the output? Use the DFT to check.

EOCP 5.12

Consider the signals

$$x(t) = \sin(2\pi t) + \cos\left(\left(\frac{3\pi}{4}\right)t\right)$$

$$x(t) = e^{-10t}\sin(2\pi t) + \cos\left(\frac{3}{4}t\right)$$

1. Are the signals periodic?
2. Find the Fourier series coefficient and/or the Fourier transform (approxima-tion) using the DFT for the signals.
3. What is the average power/total energy in the signals $x(t)$?

EOCP 5.13

Consider the signal

$$x(t) = e^{-t/10}u(t)$$

1. Find the total energy in the signal using the DFT.
2. If $x(t)$ is multiplied by $\sin(t)$, what would be the approximation to the total energy in the signal using the DFT.

EOCP 5.14

Consider the signal

$$x(t) = e^{-10t}\sin(t)u(t)$$

as an input to the system

$$h(t) = e^{-4t}u(t)$$

1. Find $y(n)$ using convolution and the DFT.
2. Subdivide the input signal into blocks and use block filtering to find $y(n)$ again.
3. Compare the results in part 1 and part 2.
4. Find the total energy in both signals using the DFT.

EOCP 5.15

Consider the signal

$$x(t) = \cos(2\pi(300)t)$$

1. Find the approximation to the Fourier transform of $x(t)$.
2. Use the Hamming windowing method and repeat part 1.
3. Use the Hanning windowing method and repeat part 1.
4. Comment on the results.

EOCP 5.16

Consider the signal

$$x(t) = 10\sin(2\pi(700)t) + \sin(2\pi(300)t)$$

1. Choose an f_s and a suitable N so that only the 700 and the 300 Hz will appear in the DFT magnitude plot. Plot the DFT magnitude.

2. Choose f_s (you should satisfy the sampling Nyquist rate in this part, too) and a suitable N so that the frequencies in $x(t)$ will appear distorted on the DFT magnitude plot. Plot the DFT magnitude.
3. How can you get a more accurate plot in part 2 to suppress the frequency distortion? Plot the DFT magnitude.

EOCP 5.17

A continuous signal has a duration of 2 s and is sampled at 64 equally spaced values.

1. What is the distance between successive frequency points?
2. What is the highest frequency in the spectrum?
3. What is the maximum frequency in the continuous signal if there is to be no aliasing?

6 State-Space and Discrete Systems

6.1 INTRODUCTION

As the interest in many scientific fields increased, modeling systems using linear time-invariant equations and tools such as transfer functions were not adequate. The state-space approach is superior to other classical methods of modeling. This modern approach can handle systems with nonzero initial conditions (modeling using transfer functions requires that initial conditions be set to zero) as well as time-variant systems. It can also handle linear and nonlinear systems. We also have been considering systems with single-input single-output. The state-space approach can handle multiple-input multiple-output systems.

Systems can have many variables. An example is an electrical circuit where the variables are the inductor current, the capacitor voltage, and the resistor voltage among others. With the state-space approach, we can solve for a selected set of these variables. The other variables in the circuit system can be found using the solution for the selected variables.

Using the state-space approach, we will follow the subsequent procedure. We will select specific variables in the system and call them state variables. No state variable selected can be written as a linear combination of the other state variables. Linear combination means that if

$$v_1(t) = 3v_3(t) + 2v_2(t) \tag{6.1}$$

where $v_1(t)$, $v_2(t)$, and $v_3(t)$ are state variables, we say that $v_1(t)$ is a linear combination of $v_2(t)$ and $v_3(t)$. If we have a first-order differential or difference equation, we will have only one state variable. If the differential or the difference equation is second order, we will have only two state variables. Similarly, if we have an nth-order differential or difference equation, we will have only n state variables. Once we select or decide on the state variables in the system under consideration, we will write a set of first-order simultaneous differential or difference equations, where the right side of these equations is a function only of the state variables (no derivatives or shifts) and the inputs to the system, and the number of these equations is determined by the number of state variables selected. We will call this set the state equation set. These state equations will be solved for the selected state variables. All other variables in the system under consideration can be solved using the solutions of these selected state variables and the inputs to the system. We can use any approach we desire to solve for these selected states. The equations we write to a set of selected outputs in the system are called output equations.

The earlier discussion relates closely to continuous systems and explains the evolution of state equations from real physical systems. In many cases, we derive discrete systems from continuous systems by many means such as sampling or transformation. The equivalence of the differential equations is the difference equations where the basic unit is the delay element. From these difference equations, the states in discrete form are obtained. Some systems are inherently discrete and the difference equation is readily obtained. The concept of the state is best understood with dynamic physical systems and that is how we explain it here.

6.2 REVIEW ON MATRIX ALGEBRA

What follows is a brief review of some of the concepts and definitions we need in this chapter. We will deal with second-order systems when we deal with hand solutions. For matrices of higher dimensions, you can consult any linear algebra book.

6.2.1 DEFINITION, GENERAL TERMS, AND NOTATIONS

A matrix is a collection of elements arranged in a rectangular or square array. The size of the matrix is determined by the number of rows and the number of columns in the matrix. A matrix $\mathbf{A}$ of m rows and n column is represented as $\mathbf{A}_{m \times n}$. If $m = 1$, then $\mathbf{A}$ is a row vector and is written as $\mathbf{A}_{1 \times n}$. If $n = 1$, then $\mathbf{A}$ is a column vector and is written as $\mathbf{A}_{m \times 1}$. If $n = m$, then $\mathbf{A}$ is a square matrix and we write it as $\mathbf{A}_{n \times n}$ or $\mathbf{A}_{m \times m}$. If all elements in the matrix are zeros, we say $\mathbf{A}$ is a null matrix or a zero matrix.

6.2.2 IDENTITY MATRIX

The identity matrix is the square matrix where elements along the main diagonal are 1s and elements off the main diagonal are 0s. A two-by-two identity matrix is

$$\mathbf{I}_{2 \times 2} = \begin{pmatrix} 1 & 0 \\ 0 & 1 \end{pmatrix} \tag{6.2}$$

6.2.3 ADDING TWO MATRICES

If $\mathbf{A} = \begin{pmatrix} a & b \\ c & d \end{pmatrix}$ and $\mathbf{B} = \begin{pmatrix} e & f \\ g & h \end{pmatrix}$ then

$$\mathbf{A} + \mathbf{B} = \begin{pmatrix} a+e & b+f \\ c+g & d+h \end{pmatrix} \tag{6.3}$$

To add two matrices, they must be of the same size. If the matrices are of higher order, the procedure is the same; we add the corresponding entries.

6.2.4 SUBTRACTING TWO MATRICES

If $A = \begin{pmatrix} a & b \\ c & d \end{pmatrix}$ and $B = \begin{pmatrix} e & f \\ g & h \end{pmatrix}$ then

$$A - B = \begin{pmatrix} a-e & b-f \\ c-g & d-h \end{pmatrix} \tag{6.4}$$

To subtract two matrices, they must be of the same size. If the matrices are of higher order, the procedure is the same; we subtract the corresponding entries.

6.2.5 MULTIPLYING A MATRIX BY A CONSTANT

If $A = \begin{pmatrix} a & b \\ c & d \end{pmatrix}$ and k is any given constant, then

$$k \begin{pmatrix} a & b \\ c & d \end{pmatrix} = \begin{pmatrix} ka & kb \\ kc & kd \end{pmatrix} \tag{6.5}$$

If the matrix A is of higher order, then k is multiplied by each entry in A.

6.2.6 DETERMINANT OF A TWO-BY-TWO MATRIX

Consider the $A_{2 \times 2}$ following matrix:

$$A_{2 \times 2} = \begin{pmatrix} a & b \\ c & d \end{pmatrix}$$

The determinant of A is

$$det(A) = ad - bc \tag{6.6}$$

If $A = \begin{pmatrix} a & b \\ c & d \end{pmatrix}$, then the transpose of A is given by

$$A^T = \begin{pmatrix} a & b \\ c & d \end{pmatrix} \tag{6.7}$$

This works for higher order matrices as well, where the first column in A becomes the first row in A^T and so on.

6.2.7 INVERSE OF A MATRIX

If $\mathbf{A} = \begin{pmatrix} a & b \\ c & d \end{pmatrix}$, then the inverse of $\mathbf{A}$ is

$$\mathbf{A}^{-1} = \frac{1}{ad - bc} \begin{pmatrix} d & -b \\ -c & a \end{pmatrix} \tag{6.8}$$

and since $1/(ad - bc)$ is a constant, we can write the inverse as

$$\mathbf{A}^{-1} = \begin{pmatrix} \dfrac{d}{ad - bc} & \dfrac{-b}{ad - bc} \\ \dfrac{-c}{ad - bc} & \dfrac{a}{ad - bc} \end{pmatrix}$$

The inverse of a square matrix exists if the determinant of the matrix is not zero. Also, to find an inverse of a certain matrix, that matrix has to be square.

The aforementioned procedure for finding the inverse is only for a 2×2 matrices. For higher order matrices, the procedure is different and is found in any linear algebra book.

6.2.8 MATRIX MULTIPLICATION

We can multiply two matrices $\mathbf{A}$ and $\mathbf{B}$ if the number of columns in $\mathbf{A}$ is equal to the number of rows in $\mathbf{B}$. If $\mathbf{A}_{m \times n}$ is to be multiplied by $\mathbf{B}_{r \times p}$, then n must be equal to r and the resulting matrix should have m rows and p columns:

If

$$\mathbf{A} = \begin{pmatrix} a & b \\ c & d \end{pmatrix} \text{ and } \mathbf{B} = \begin{pmatrix} e & f \\ g & h \end{pmatrix}$$

then if we multiply $\mathbf{A}$ by $\mathbf{B}$ and let matrix $\mathbf{C}$ hold the resulting product, the size of $\mathbf{C}$ is 2×2. We could multiply $\mathbf{A}$ by $\mathbf{B}$ because the number of columns in $\mathbf{A}$, which is two, is equal to the number of rows in $\mathbf{B}$, which is also two. The multiplication of $\mathbf{A}$ by $\mathbf{B}$ is $\mathbf{C}$ and it is

$$\mathbf{C} = \mathbf{AB} = \begin{pmatrix} a & b \\ c & d \end{pmatrix}\begin{pmatrix} e & f \\ g & h \end{pmatrix} = \begin{pmatrix} ae + bg & af + bh \\ ce + dg & cf + dh \end{pmatrix} \tag{6.9}$$

We multiply the first row of $\mathbf{A}$, element by element, by all the columns of $\mathbf{B}$. Similarly, we take the second row of $\mathbf{A}$ and multiply it by all the columns of $\mathbf{B}$. Note that in general $\mathbf{AB}$ is not the same as $\mathbf{BA}$. The rules for multiplication have to be observed.

6.2.9 Eigenvalues of a Matrix

The eigenvalues of a matrix $\mathbf{A}$ are the roots of the determinant of $(\lambda \mathbf{I} - \mathbf{A})$, where $\mathbf{I}$ is the identity matrix and λ is a variable.

6.2.10 Diagonal Form of a Matrix

A matrix $\mathbf{A}$ is in diagonal form if all elements in the matrix that are off the diagonal are zeros. If $\mathbf{A}$ is not diagonal, we can make it diagonal by finding the matrix $\mathbf{P}$ that contains the eigenvectors of $\mathbf{A}$. So if $\mathbf{A}$ is not a diagonal matrix, then $(\mathbf{P}^{-1}\mathbf{A}\mathbf{P})$ will transform $\mathbf{A}$ into a diagonal matrix. If the eigenvalues of $\mathbf{A}$ are distinct (distinct means that no one eigenvalue is equal to the other), then $\mathbf{P}^{-1}\mathbf{A}\mathbf{P}$ will look like a diagonal matrix, where the elements on the main diagonal are the eigenvalues of $\mathbf{A}$. If some of the eigenvalues of $\mathbf{A}$ are repeated, then the matrix $\mathbf{P}^{-1}\mathbf{A}\mathbf{P}$ will be in a block diagonal form or what is known as Jordan form.

6.2.11 Eigenvectors of a Matrix

The eigenvectors of a matrix $\mathbf{A}$ are the nonzero roots of the homogeneous matrix equation

$$\left(\lambda_i \mathbf{I} - \mathbf{A}\right)\mathbf{p}_i = 0 \tag{6.10}$$

where $\mathbf{p}_i$ is a column vector that represents the eigenvector for a certain eigenvalue λ_i. All eigenvectors must be independent. If we have n distinct eigenvalues, then we will have n independent eigenvectors for each eigenvalue, and each eigenvector is obtained as the nonzero solution to

$$\left(\lambda_i \mathbf{I} - \mathbf{A}\right)\mathbf{p}_i = 0.$$

If $\mathbf{A}$ is $n \times n$ and we have $n - k$ distinct eigenvalues (again distinct means no one eigenvalue is equal to any of the remaining $n - 1$ eigenvalues), then we will have $n - k$ independent eigenvectors for each distinct eigenvalue and each eigenvector is obtained as the nonzero solution of $(\lambda_i \mathbf{I} - \mathbf{A})\mathbf{p}_i = \mathbf{0}$. Each remaining eigenvalue in the set k of eigenvalues is therefore not distinct. We may have groups of repeated eigenvalues in the k set of eigenvalues. We will assume here that we have only one set, which is k and thus we have k repeated eigenvalues. The following procedure can be applied to other repeated sets in the set k.

Assume that $k = 3$. Let the repeated eigenvalues be denoted as λ_r and let us denote the eigenvectors as $\mathbf{p}_{1r}$, $\mathbf{p}_{2r}$, and $\mathbf{p}_{3r}$.

To find $\mathbf{p}_{1r}$, we will find the nonzero solution to $(\lambda_r \mathbf{I} - \mathbf{A})\mathbf{p}_{1r} = \mathbf{0}$.
To find $\mathbf{p}_{2r}$, we will find the nonzero solution to $(\lambda_r \mathbf{I} - \mathbf{A})\mathbf{p}_{2r} = \mathbf{p}_{1r}$.
To find $\mathbf{p}_{3r}$, we will find the nonzero solution to $(\lambda_r \mathbf{I} - \mathbf{A})\mathbf{p}_{3r} = \mathbf{p}_{2r}$.

These are called generalized eigenvectors corresponding to the three repeated eigenvalues. In some situations, even if we have repeated eigenvalues, we still can get independent eigenvectors if the matrix $\mathbf{A}$ is symmetric. We do that by finding a nonzero solution to $(\lambda_i \mathbf{I} - \mathbf{A})\mathbf{p}_i = \mathbf{0}$, where λ_i is the repeated eigenvalue.

6.3   GENERAL REPRESENTATION OF SYSTEMS IN STATE SPACE

Thus far, we have seen many representations of discrete linear time-invariant systems. We have seen the difference equation representation, the block diagram representation, the z-transform representation, and the impulse response representation. Given any representation we should be able to deduce the other. In this section, we will study the state-space representation starting with the difference equation. Using state-space representation, a difference equation of order n can be reduced to n first-order difference equations.

6.3.1   RECURSIVE SYSTEMS

Consider the following fourth-order equation:

$$y(n) - a_1 y(n-1) - a_2 y(n-2) - a_3 y(n-3) - a_4 y(n-4) = b_0 x(n)$$

We will have four states v_1 through v_4 since we have a fourth-order difference equation. Let

$$v_1(n) = y(n-4)$$

$$v_2(n) = y(n-3)$$

$$v_3(n) = y(n-2)$$

$$v_4(n) = y(n-1)$$

from which we write

$$v_1(n+1) = y(n-4+1) = y(n-3) v_2(n)$$

$$v_2(n+1) = y(n-3+1) = y(n-2) v_3(n)$$

$$v_3(n+1) = y(n-2+1) = y(n-1) v_4(n)$$

and we can also write the output as

$$y(n) = a_1 y(n-1) - a_2 y(n-2) - a_3 y(n-3) - a_4 y(n-4) = b_0 x(n)$$

$$= a_1 v_4(n) + a_2 v_3(n) + a_3 v_2(n) + a_4 v_1(n) + b_0 x(n)$$

Therefore, we can finally write

$$v_1(n+1) = v_2(n)$$

$$v_2(n+1) = v_3(n)$$

$$v_3(n+1) = v_4(n)$$

$$v_4(n+1) = a_4 v_1(n) + a_3 v_2(n) + a_2 v_3(n) + a_1 v_4(n) + b_0 x(n)$$

Notice that the right side of all of these four state equations is a function of the states $v_1(n)$ through $v_4(n)$ and the input. This should be the case. No term such as $v_1(n-1)$ or $v_2(n-3)$ should appear on the right side of these equations.

We can now arrange the state equations in matrix form and write

$$
\begin{pmatrix} v_1(n+1) \\ v_2(n+1) \\ v_3(n+1) \\ v_4(n+1) \end{pmatrix} = \begin{pmatrix} 0 & 1 & 0 & 0 \\ 0 & 0 & 1 & 0 \\ 0 & 0 & 0 & 1 \\ a_4 & a_3 & a_2 & a_1 \end{pmatrix} \begin{pmatrix} v_1(n) \\ v_2(n) \\ v_3(n) \\ v_4(n) \end{pmatrix} + \begin{pmatrix} 0 \\ 0 \\ 0 \\ b_0 \end{pmatrix} x(n)
$$

and for the output equation, we have

$$
y(n) = \begin{pmatrix} a_4 & a_3 & a_2 & a_1 \end{pmatrix} \begin{pmatrix} v_1(n) \\ v_2(n) \\ v_3(n) \\ v_4(n) \end{pmatrix} + b_0 x(n)
$$

with

$$
\mathbf{A} = \begin{pmatrix} 0 & 1 & 0 & 0 \\ 0 & 0 & 1 & 0 \\ 0 & 0 & 0 & 1 \\ a_4 & a_3 & a_2 & a_1 \end{pmatrix} \quad \mathbf{B} = \begin{pmatrix} 0 \\ 0 \\ 0 \\ b_0 \end{pmatrix} \quad \mathbf{C} = \begin{pmatrix} a_4 & a_3 & a_2 & a_1 \end{pmatrix} \ \mathbf{D} = \begin{pmatrix} b_0 \end{pmatrix}
$$

We can write the state and output equations in matrix form as

$$
\mathbf{v}(n+1) = \mathbf{A}\mathbf{v}(n) + \mathbf{B}\mathbf{x}(n)
$$

$$
\mathbf{y}(n) = \mathbf{C}\mathbf{v}(n) + \mathbf{D}\mathbf{x}(n)
$$

When $\mathbf{v}(n+1)$ is a 4×1 matrix, $\mathbf{A}$ is 4×4 matrix, $\mathbf{B}$ is 4×1 matrix, $\mathbf{C}$ is a 1×4 matrix, and $\mathbf{D}$ is a 1×1 matrix. Notice that these state equations are now in the form of a first-order matrix difference equation.

6.3.2 NONRECURSIVE SYSTEMS

Consider the following difference equation:

$$
y(n) = a_0 x(n) + a_1 x(n-1) + a_2 x(n-2) + a_3 x(n-3)
$$

Let

$$v_1(n) = x(n-3)$$

$$v_2(n) = x(n-2)$$

$$v_3(n) = x(n-1)$$

Then

$$v_1(n+1) = x(n-2) = v_2(n)$$

$$v_2(n+1) = x(n-1) = v_3(n)$$

$$v_3(n+1) = x(n)$$

and the output equation is

$$y(n) = a_0 x(n) + a_1 x(n-1) + a_2 x(n-2) + a_3 x(n-3)$$

Substituting the states in this output equation gives

$$y(n) = a_0 x(n) + a_1 v_3(n) + a_2 v_2(n) + a_3 v_1(n)$$

The state matrix equations are then

$$\begin{pmatrix} v_1(n+1) \\ v_2(n+1) \\ v_3(n+1) \end{pmatrix} = \begin{pmatrix} 0 & 1 & 0 \\ 0 & 0 & 1 \\ 0 & 0 & 0 \end{pmatrix} \begin{pmatrix} v_1(n) \\ v_2(n) \\ v_3(n) \end{pmatrix} + \begin{pmatrix} 0 \\ 0 \\ 1 \end{pmatrix} x(n)$$

$$y(n) = \begin{pmatrix} a_3 & a_2 & a_1 \end{pmatrix} \begin{pmatrix} v_1(n) \\ v_2(n) \\ v_3(n) \end{pmatrix} + a_0 x(n)$$

where in this case,

$$A = \begin{pmatrix} 0 & 1 & 0 \\ 0 & 0 & 1 \\ 0 & 0 & 0 \end{pmatrix} \quad B = \begin{pmatrix} 0 \\ 0 \\ 1 \end{pmatrix} \quad C = \begin{pmatrix} a_3 & a_2 & a_1 \end{pmatrix} \quad D = (a_0)$$

6.3.3 FROM THE BLOCK DIAGRAM TO STATE SPACE

Consider the block diagram shown in Figure 6.1. To obtain the states from the block diagram, we need first to determine the order of the system so that we know how many states to consider. Then we will let the output of every delay represent a state. Since the system is second order, we will have two states $v_1(n)$ and $v_2(n)$. Let the output of the first delay from the input side be denoted as $v_1(n)$. Then the input of this delay is $v_1(n + 1)$. Therefore, from the graph, we have

$$v_1(n+1) = x(n) + a_0 y(n)$$

Let the output of the second delay be $v_2(n)$. Then

$$v_1(n+1) = x(n) + a_0 \left[b_0 x(n) + v_2(n) \right]$$

or

$$v_1(n+1) = \left[1 + a_0 b_0 \right] x(n) + a_0 v_2(n)$$

The input to the second delay is $v_2(n + 1)$ and is

$$v_2(n+1) = v_1(n) + a_1 y(n) = v_1(n) + a_1 \left[b_0 x(n) + v_2(n) \right]$$

$$= v_1(n) + a_1 b_0 x(n) + a_1 v_2(n)$$

Also, the output $y(n)$ is

$$y(n) = b_0 x(n) + v_2(n)$$

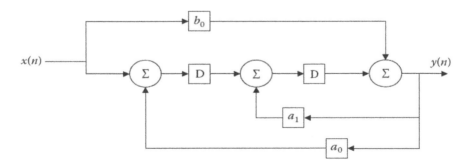

FIGURE 6.1 Block diagram representation 1.

Therefore, the state and the output matrix equations are

$$\begin{pmatrix} v_1(n+1) \\ v_2(n+1) \end{pmatrix} = \begin{pmatrix} 0 & a_0 \\ 1 & a_1 \end{pmatrix} \begin{pmatrix} v_1(n) \\ v_2(n) \end{pmatrix} + \begin{pmatrix} 1+a_0b_0 \\ a_1b_0 \end{pmatrix} x(n)$$

$$\mathbf{y}(n) = \begin{pmatrix} 0 & 1 \end{pmatrix} \begin{pmatrix} v_1(n) \\ v_2(n) \end{pmatrix} + b_0\mathbf{x}(n)$$

where

$$\mathbf{A} = \begin{pmatrix} 0 & a_0 \\ 1 & a_1 \end{pmatrix} \quad \mathbf{B} = \begin{pmatrix} 1+a_0b_0 \\ a_1b_0 \end{pmatrix} \quad \mathbf{C} = \begin{pmatrix} 0 & 1 \end{pmatrix} \quad \mathbf{D} = (b_0)$$

Consider another block diagram shown in Figure 6.2. This system is third order and must have three states. Let the output of the first delay from the input side be $v_1(n)$, the output of the second delay be $v_2(n)$, and the output of the last delay be $v_3(n)$. The input of the first delay is

$$v_1(n+1) = x(n) + a_0v_1(n)$$

The input of the second delay is

$$v_2(n+1) = v_1(n) + a_1v_2(n)$$

The input of the third delay is

$$v_3(n+1) = v_2(n) + a_2v_3(n)$$

The output $y(n)$ is

$$y(n) = v_3(n)$$

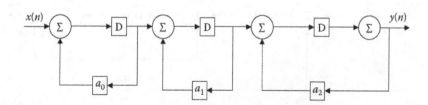

FIGURE 6.2   Block diagram representation 2.

We then have the state-space system in matrix form as

$$
\begin{pmatrix} v_1(n+1) \\ v_2(n+1) \\ v_3(n+1) \end{pmatrix} = \begin{pmatrix} a_0 & 0 & 0 \\ 1 & a_1 & 0 \\ 0 & 0 & a_2 \end{pmatrix} \begin{pmatrix} v_1(n) \\ v_2(n) \\ v_3(n) \end{pmatrix} + \begin{pmatrix} 1 \\ 0 \\ 0 \end{pmatrix} x(n)
$$

$$
y(n) = \begin{pmatrix} 0 & 0 & 1 \end{pmatrix} \begin{pmatrix} v_1(n) \\ v_2(n) \\ v_3(n) \end{pmatrix} + (0)x(n)
$$

where

$$
A = \begin{pmatrix} a_0 & 0 & 0 \\ 1 & a_1 & 0 \\ 0 & 0 & a_2 \end{pmatrix} \quad B = \begin{pmatrix} 1 \\ 0 \\ 0 \end{pmatrix} \quad C = \begin{pmatrix} 0 & 0 & 1 \end{pmatrix} \quad D = (0)
$$

6.3.4 FROM THE TRANSFER FUNCTION $H(z)$ TO STATE SPACE

Given $H(z)$, the transfer function is the z-domain. We can obtain the state-space realization in many ways. We will illustrate that using an example.

Example 6.1

Consider the following system:

$$
H(z) = \frac{z-4}{z^2 + 5z + 6}
$$

What is the state-space representation?

SOLUTION

1. Using direct realization, the block diagram for $H(z)$ is shown in Figure 6.3. Let the output of the first delay be $v_1(n)$. Then,

$$
v_1(n+1) = -4x(n) - 6y(n) = -4x(n) - 6\left[v_2(n)\right]
$$

$$
-6v_2(n) - 4x(n)
$$

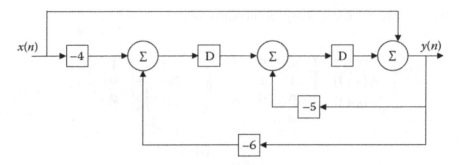

FIGURE 6.3 Block diagram for Example 6.1.

where $v_2(n)$ is the output of the second delay. Also,

$$v_2(n+1) = v_1(n) - 5v_2(n) = v_1(n) - 5\big[x(n) + v_2(n)\big]$$

$$= v_1(n) - 5v_2(n)$$

The output is

$$y(n) = v_2(n)$$

Then the state-space representation is

$$\begin{pmatrix} v_1(n+1) \\ v_2(n+1) \end{pmatrix} = \begin{pmatrix} 0 & -6 \\ 1 & -5 \end{pmatrix} \begin{pmatrix} v_1(n) \\ v_2(n) \end{pmatrix} + \begin{pmatrix} -4 \\ 0 \end{pmatrix} x(n)$$

$$y(n) = \begin{pmatrix} 0 & 1 \end{pmatrix} \begin{pmatrix} v_1(n) \\ v_2(n) \end{pmatrix} + (0)\, x(n)$$

With

$$\mathbf{A} = \begin{pmatrix} 0 & -6 \\ 1 & -5 \end{pmatrix}\ \mathbf{B} = \begin{pmatrix} -4 \\ 0 \end{pmatrix}\ \mathbf{C} = \begin{pmatrix} 0 & 1 \end{pmatrix}\ \mathbf{D} = (0)$$

2. The transfer function can also be written as

$$H(z) = \frac{z-4}{z^2 + 2\,z + 3}\frac{1}{}$$

This is what we call the cascade system realization, and the block diagram can be drawn as in Figure 6.4. Again the system is still second order

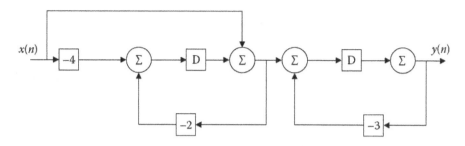

FIGURE 6.4   Block diagram for Example 6.1 in cascade form.

and if we let the output of the first delay be $v_1(n)$ and the output of the second delay be $v_2(n)$, then

$$v_1(n+1) = -4x(n) - 2x(n) - 2v_1(n)$$

$$v_2(n+1) = x(n) + v_1(n) - 3y(n)$$

$$= x(n) + v_1(n) - 3\left[v_2(n)\right]$$

The output is

$$y(n) = v_2(n)$$

We can now clean the previous state equations and write

$$v_1(n+1) = -2v_1(n) - 6x(n)$$

$$v_2(n+1) = v_1(n) - 3v_2(n) + 1x(n)$$

$$y(n) = v_2(n)$$

with

$$\begin{pmatrix} v_1(n+1) \\ v_2(n+1) \end{pmatrix} = \begin{pmatrix} -2 & 0 \\ -1 & -3 \end{pmatrix} \begin{pmatrix} v_1(n) \\ v_2(n) \end{pmatrix} + \begin{pmatrix} -6 \\ -1 \end{pmatrix} x(n)$$

$$y(n) = \begin{pmatrix} 0 & 1 \end{pmatrix} \begin{pmatrix} v_1(n) \\ v_2(n) \end{pmatrix} + (0)x(n)$$

Where

$$A = \begin{pmatrix} -2 & 0 \\ -1 & -3 \end{pmatrix} \quad B = \begin{pmatrix} -6 \\ -1 \end{pmatrix} \quad C = \begin{pmatrix} 0 & 1 \end{pmatrix} \quad D = (0)$$

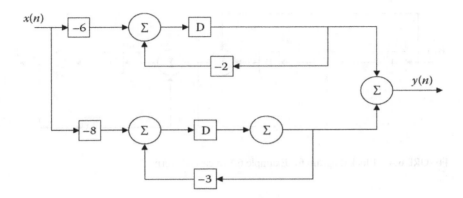

FIGURE 6.5 Block diagram for Example 6.1 in parallel form.

3. We can also write $H(z)$ in a parallel realization form as

$$H(z) = \frac{-6}{z+2} \frac{-7}{z+3}$$

The block diagram is shown in Figure 6.5. Let $v_1(n)$ be the output of the upper delay and $v_2(n)$ be the output of the lower delay. Then

$$v_1(n+1) = -6x(n) - 2v_1(n)$$

and

$$v_2(n+1) = -7x(n) - 3v_2(n)$$

The output is

$$y(n) = v_1(n) + v_2(n)$$

The state-space system is then

$$\begin{pmatrix} v_1(n+1) \\ v_2(n+1) \end{pmatrix} = \begin{pmatrix} -2 & 0 \\ 0 & -3 \end{pmatrix} \begin{pmatrix} v_1(n) \\ v_2(n) \end{pmatrix} + \begin{pmatrix} -6 \\ 7 \end{pmatrix} x(n)$$

$$y(n) = \begin{pmatrix} 1 & 1 \end{pmatrix} \begin{pmatrix} v_1(n) \\ v_2(n) \end{pmatrix} + (0)x(n)$$

with

$$A = \begin{pmatrix} -2 & 0 \\ 0 & -3 \end{pmatrix} \quad B = \begin{pmatrix} -6 \\ 7 \end{pmatrix} \quad C = \begin{pmatrix} 1 & 1 \end{pmatrix} \quad D = (0)$$

4. Let us now draw a simulation diagram from

$$H(z) = \frac{z-4}{z^2+5z+6}$$

that is different from the diagram we drew earlier in this example. From

$$H(z) = \frac{Y(z)}{X(z)}$$

we have

$$Y(z)\left[z^2+5z+6\right] = X(z)\left[z-4\right]$$

By taking the inverse z-transform, we get the following difference equation:

$$6y(n) + 5y(n+1) + y(n+2) = -4x(n) + x(n+1)$$

This is a second-order difference equation and its block diagram is shown in Figure 6.6. Let the output of the first delay on the top be $v_2(n)$ and the output of the lower delay be $v_1(n)$. Then we have the state and the output equations as

$$v_2(n+1) = x(n) - 5v_2(n) - 6v_1(n)$$

$$v_1(n+1) = v_2(n)$$

$$y(n) = v_2(n) - 4v_1(n)$$

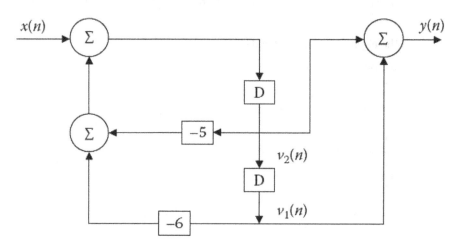

FIGURE 6.6 Block diagram for Example 6.1.

Then the matrix state-space representation is

$$
\begin{pmatrix} v_1(n+1) \\ v_2(n+1) \end{pmatrix} = \begin{pmatrix} 0 & 1 \\ -6 & -5 \end{pmatrix} \begin{pmatrix} v_1(n) \\ v_2(n) \end{pmatrix} + \begin{pmatrix} 0 \\ 1 \end{pmatrix} x(n)
$$

$$
y = \begin{pmatrix} -4 & 1 \end{pmatrix} \begin{pmatrix} v_1(n) \\ v_2(n) \end{pmatrix} + (0) x(n)
$$

with

$$
A = \begin{pmatrix} 0 & 1 \\ -6 & -5 \end{pmatrix} \quad B = \begin{pmatrix} 0 \\ 1 \end{pmatrix} \quad C = \begin{pmatrix} 0 & -4 & 1 \end{pmatrix} \quad D = (0)
$$

Notice that in all of these four state-space realizations we had different **A**, **B**, **C**, and **D** matrices. To check that each system is a true realization, the eigenvalues in each case should be the same. In case 1,

$$
A = \begin{pmatrix} 0 & -6 \\ 1 & -5 \end{pmatrix}
$$

and the eigenvalues for the system are the roots of the determinant of $(\lambda I - A)$.

$$
\lambda \begin{pmatrix} 1 & 0 \\ 0 & 1 \end{pmatrix} - \begin{pmatrix} 0 & -6 \\ 1 & -5 \end{pmatrix} = \begin{pmatrix} \lambda & 6 \\ -1 & \lambda+5 \end{pmatrix}
$$

$\det(\lambda I - A) = \lambda^2 + 5\lambda + 6$ and the eigenvalues are at -2 and -3.

In case 2

$$
A = \begin{pmatrix} -2 & 0 \\ 1 & -3 \end{pmatrix} \text{ and } \det(\lambda I - A) = \det \begin{pmatrix} \lambda+2 & 0 \\ 1 & \lambda+3 \end{pmatrix} = \lambda^2 + 5\lambda + 6
$$

The eigenvalues are at -2 and -3.

In case 3,

$$
A = \begin{pmatrix} -2 & 0 \\ 0 & -3 \end{pmatrix} \text{ and } \det(\lambda I - A) = \det \begin{pmatrix} \lambda+2 & 0 \\ 1 & \lambda+3 \end{pmatrix} = \lambda^2 + 5\lambda + 6
$$

The eigenvalues are at -2 and -3.

In the last case,

$$
A = \begin{pmatrix} 0 & 1 \\ -6 & -5 \end{pmatrix} \text{ and } \det(\lambda I - A) = \lambda^2 + 5\lambda + 6
$$

The eigenvalues are at -2 and -3.

Example 6.2

Consider the system

$$\frac{b_0 z^5 + b_1 z^4 + b_2 z^3 + b_3 z^2 + b_4 z + b_5}{z^5 + a_1 z^4 + a_2 z^3 + a_3 z^2 + a_4 z + a_5}$$

with the coefficient of z^5 in the denominator always unity. What is the state-space representation?

SOLUTION

We start by the block diagram shown in Figure 6.7. From the graph we see that the state equations are

$$v_1(n+1) = v_2(n)$$

$$v_2(n+1) = v_3(n)$$

$$v_3(n+1) = v_4(n)$$

$$v_4(n+1) = v_5(n)$$

$$v_5(n+1) = x(n) - a_1 v_5(n) - a_2 v_4(n) - a_3 v_3(n) - a_4 v_2(n) - a_5 v_1(n)$$

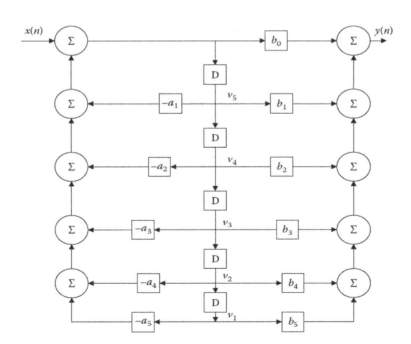

FIGURE 6.7 Block diagram for Example 6.2.

and the output equation is

$$y(n) = (b_5 - a_5 b_0) v_1(n) + (b_4 - a_4 b_0) v_2(n) + (b_3 - a_3 b_0) v_3(n)$$
$$+ (b_2 - a_2 b_0) v_4(n) + (b_1 - a_1 b_0) v_5(n) + b_0 x(n)$$

The state equations and the output equation in state-space matrix form are

$$
\begin{pmatrix} v_1(n+1) \\ v_2(n+1) \\ v_3(n+1) \\ v_4(n+1) \\ v_5(n+1) \end{pmatrix}
=
\begin{pmatrix}
0 & 1 & 0 & 0 & 0 \\
0 & 0 & 1 & 0 & 0 \\
0 & 0 & 0 & 1 & 0 \\
0 & 0 & 0 & 0 & 1 \\
-a_5 & -a_4 & -a_3 & -a_2 & -a_1
\end{pmatrix}
\begin{pmatrix} v_1(n) \\ v_2(n) \\ v_3(n) \\ v_4(n) \\ v_5(n) \end{pmatrix}
+
\begin{pmatrix} 0 \\ 0 \\ 0 \\ 0 \\ 1 \end{pmatrix} x(n)
$$

$$
y(n) = \left[(b_5 - a_5 b_0)(b_4 - a_4 b_0)(b_3 - a_3 b_0)(b_2 - a_2 b_0)(b_1 - a_1 b_0) \right]
\begin{pmatrix} v_1(n) \\ v_2(n) \\ v_3(n) \\ v_4(n) \\ v_5(n) \end{pmatrix}
+ (b_0) x(n)
$$

Example 6.3

Given the system

$$H(z) = \frac{2z^2 + 3z^2 + 4z + 8}{z^2 + 5z^2 + 6z + 10}$$

What is the state-space representation? Find **A**, **B**, **C**, and **D**.

SOLUTION

From Example 6.2, we can write these matrices by inspection and get

$$
A = \begin{pmatrix} 0 & 1 & 0 \\ 0 & 0 & 1 \\ -10 & -6 & -5 \end{pmatrix}
\quad
B = \begin{pmatrix} 0 \\ 0 \\ 1 \end{pmatrix}
$$

$$
C = \begin{pmatrix} 8-10(2) & 4-6(2) & 3-5(2) \end{pmatrix} = \begin{pmatrix} -12 & -8 & -7 \end{pmatrix} \quad D = (2)
$$

6.4 SOLUTION OF THE STATE-SPACE EQUATIONS IN THE z-DOMAIN

We are given the state-space system

$$v(n+1) = \mathbf{A}v(n) + \mathbf{B}\mathbf{x}(n)$$

$$\mathbf{y}(n) = \mathbf{C}v(n) + \mathbf{D}\mathbf{x}(n)$$

By taking the z-transform on the previous state equations, we will get

$$z\mathbf{V}(z) - z\mathbf{v}(0) = \mathbf{A}\mathbf{V}(z) + \mathbf{B}\mathbf{X}(z)$$

Notice that z is a scalar and cannot be subtracted from $\mathbf{A}$. Therefore, we write the previous equations as

$$(z\mathbf{I} - \mathbf{A})\mathbf{V}(z) = z\mathbf{v}(0) + \mathbf{B}\mathbf{X}(z)$$

where $\mathbf{I}$ is the identity matrix. Solving the previous equation for the states, we get

$$\mathbf{V}(z) = z(z\mathbf{I} - \mathbf{A})^{-1}\mathbf{v}(0) + (z\mathbf{I} - \mathbf{A})^{-1}\mathbf{B}\mathbf{X}(z) \qquad (6.11)$$

$v(n)$ is the inverse transform of the previous equation. For the output y(n), we have

$$\mathbf{Y}(z) = \mathbf{C}\mathbf{V}(z) + \mathbf{D}\mathbf{X}(z)$$

$$= \mathbf{C}\left[z(z\mathbf{I} - \mathbf{A})^{-1}\mathbf{v}(0) + (z\mathbf{I} - \mathbf{A})^{-1}\mathbf{B}\mathbf{X}(z)\right] + \mathbf{D}\mathbf{X}(z)$$

This is a good place to try to find $H(z)$ from the state-space equations. With $v(0) = 0$, we have

$$\mathbf{Y}(z) = \mathbf{C}\mathbf{V}(z) + \mathbf{D}\mathbf{X}(z)$$

$$\frac{\mathbf{Y}(z)}{\mathbf{X}(z)} = \left[\mathbf{C}(z\mathbf{I} - \mathbf{A})^{-1}\mathbf{B} + \mathbf{D}\right]$$

or

$$\frac{\mathbf{Y}(z)}{\mathbf{X}(z)} = \left[\mathbf{C}(z\mathbf{I} - \mathbf{A})^{-1}\mathbf{B} + \mathbf{D}\right] \qquad (6.12)$$

6.5 GENERAL SOLUTION OF THE STATE EQUATION IN REAL TIME

The state equation in matrix form is repeated here

$$v(n+1) = \mathbf{A}v(n) + \mathbf{B}\mathbf{x}(n)$$

Assume that the initial condition vector $\mathbf{v}(0)$ is known for $n = 0$. For $n = 0$, the state equation becomes

$$\mathbf{v}(1) = \mathbf{A}\mathbf{v}(0) + \mathbf{B}\mathbf{x}(0)$$

For $n = 1$, the state equation is

$$\mathbf{v}(2) = \mathbf{A}\mathbf{v}(1) + \mathbf{B}\mathbf{x}(1)$$

If we substitute $\mathbf{v}(1)$ in the equation for $\mathbf{v}(2)$, we get

$$\mathbf{A}\left[\mathbf{A}\mathbf{v}(0) + \mathbf{B}\mathbf{x}(0)\right] + \mathbf{B}\mathbf{x}(1) = \mathbf{A}^2\mathbf{v}(0) + \mathbf{A}\mathbf{B}\mathbf{x}(0) + \mathbf{B}\mathbf{x}(1)$$

For $n = 2$, the state equation is

$$\mathbf{v}(3) = \mathbf{A}\mathbf{v}(2) + \mathbf{B}\mathbf{x}(2) = \mathbf{A}\left[\mathbf{A}^2\mathbf{v}(0) + \mathbf{A}\mathbf{B}\mathbf{x}(0) + \mathbf{B}\mathbf{x}(1)\right] + \mathbf{B}\mathbf{x}(2)$$

$$= \mathbf{A}^3\mathbf{v}(0) + \mathbf{A}^2\mathbf{B}\mathbf{x}(0) + \mathbf{A}\mathbf{B}\mathbf{x}(1) + \mathbf{B}\mathbf{x}(2)$$

From the previous equation, we deduce that for $n \geq 0$ the states are given by

$$\mathbf{v}(n) = \mathbf{A}^n\mathbf{v}(0) + \sum_{k=0}^{n-1} \mathbf{A}^{(n-1-k)}\mathbf{B}\mathbf{x}(k) \tag{6.13}$$

Let $\mathbf{A}^n = \boldsymbol{\varphi}(n)$ the state transition matrix. Then the solution for $\mathbf{v}(n)$ becomes

$$\mathbf{v}(n) = \boldsymbol{\varphi}(n)\mathbf{v}(0) + \sum_{k-0}^{n-1} \boldsymbol{\varphi}(n-1-k)\mathbf{B}\mathbf{x}(k) \tag{6.14}$$

$$\mathbf{y}(n) = \mathbf{C}\boldsymbol{\varphi}(n)\mathbf{v}(0) + \sum_{k=0}^{n-1} \mathbf{C}\boldsymbol{\varphi}(n-1-k)\mathbf{B}\mathbf{x}(k) + \mathbf{D}\mathbf{x}(n) \tag{6.15}$$

In general, we do not attempt to solve for $\mathbf{A}^n$ as a function of n. But we use recursion to solve for $\mathbf{v}(n)$.

Now let us compare the solutions for the state vector using the time domain given by

$$\mathbf{v}(n) = \boldsymbol{\varphi}(n)\mathbf{v}(0) + \sum_{k=0}^{n-1} \boldsymbol{\varphi}(n-1-k)\mathbf{B}\mathbf{x}(k)$$

and the solution using the z-domain given as

$$\mathbf{V}(z) = z(z\mathbf{I} - \mathbf{A})^{-1}\mathbf{v}(0) + (z\mathbf{I} - \mathbf{A})^{-1}\mathbf{B}\mathbf{X}(z)$$

A close look at these solutions reveals that the transition matrix is found by comparing the coefficients of $\mathbf{v}(0)$ and it is the inverse transform of

$$z(z\mathbf{I} - \mathbf{A})^{-1} \qquad (6.16)$$

6.6 PROPERTIES OF A^N AND ITS EVALUATION

In real time, we will use the computer to solve for the states $\mathbf{v}(n)$. We can also use the z-transform method to find a closed-form solution for the states if the system order is reasonably low. The transition matrix $\mathbf{A}^n$ has some interesting properties.

1. The response to a system with zero input is

$$\mathbf{v}(n) = \phi(n)\mathbf{v}(0)$$

For $n = 0$, $\mathbf{v}(0) = \phi(0)\mathbf{v}(0)$. This indicates that $\phi(0) = \mathbf{I}$, where $\mathbf{I}$ is the identity matrix. This result is useful if we want to confirm the correctness of $\boldsymbol{\phi}(n)$.

2. With $\boldsymbol{\phi}(n) = \mathbf{A}^n$, let $n = n_1 + n_2$ to write

$$\phi(n_1 + n_2) = \mathbf{A}^{n_1+n_2} = \mathbf{A}^{n_1}\mathbf{A}^{n_2} = \phi(n_1)\phi(n_2)$$

3. With $\boldsymbol{\phi}(n) = \mathbf{A}^n$, let $n = -m$ to get

$$\phi(-m) = \mathbf{A}^{-m} = \left(\mathbf{A}^m\right)^{-1} = \phi^{-1}(m)$$

Thus, we can write

$$\phi^{-1}(n) = \phi(-n)$$

Example 6.4

Consider the following system:

$$\mathbf{v}(n+1) = \begin{pmatrix} -1 & 0 \\ 0 & -2 \end{pmatrix} \mathbf{v}(n) + \begin{pmatrix} 0 \\ 1 \end{pmatrix} \mathbf{x}(n)$$

$$\mathbf{y}(n) = \begin{pmatrix} 0 & 1 \end{pmatrix} \mathbf{v}(n)$$

Find the states and the output, $\mathbf{v}(n)$ and $\mathbf{y}(n)$.

Let $\mathbf{v}(0) = \begin{pmatrix} 0 \\ 1 \end{pmatrix}$ with $\mathbf{x}(n) = 0$.

SOLUTION

In real time, we can use recursion to solve for $\mathbf{v}(n)$ and $\mathbf{y}(n)$. The state solution is

$$\mathbf{v}(n) = \mathbf{A}^n \mathbf{v}(0) + \sum_{k=0}^{n-1} \mathbf{A}^{(n-1-k)} \mathbf{B} \mathbf{x}(k)$$

But with $\mathbf{x}(n) = 0$,

$$\mathbf{v}(n) = \mathbf{A}^n \mathbf{v}(0)$$

Now we start the iteration and write

$$\mathbf{v}(1) = \mathbf{A}\mathbf{v}(0) = \begin{pmatrix} -1 & 0 \\ 0 & -2 \end{pmatrix} \begin{pmatrix} 0 \\ 1 \end{pmatrix} = \begin{pmatrix} 0 \\ -2 \end{pmatrix}$$

$$\mathbf{v}(2) = \mathbf{A}\mathbf{v}(1) = \begin{pmatrix} -1 & 0 \\ 0 & -2 \end{pmatrix} \begin{pmatrix} 0 \\ -2 \end{pmatrix} = \begin{pmatrix} 0 \\ 4 \end{pmatrix}$$

$$\mathbf{v}(3)\mathbf{A}\mathbf{v}(2) = \begin{pmatrix} -1 & 0 \\ 0 & -2 \end{pmatrix} \begin{pmatrix} 0 \\ 4 \end{pmatrix} = \begin{pmatrix} 0 \\ -8 \end{pmatrix}$$

and so on. But since $\mathbf{A}$ is in a diagonal form, we have

$$\mathbf{A}^n = \begin{pmatrix} (-1)^n & 0 \\ 0 & (-2)^n \end{pmatrix}$$

and therefore the state solution is

$$\mathbf{v}(n) = \mathbf{A}^n \begin{pmatrix} 0 \\ 1 \end{pmatrix}$$

Thus,

$$\mathbf{v}(0) = \mathbf{A}^0 \begin{pmatrix} 0 \\ 1 \end{pmatrix} = \begin{pmatrix} 1 & 0 \\ 0 & 1 \end{pmatrix} \begin{pmatrix} 0 \\ 1 \end{pmatrix} = \begin{pmatrix} 0 \\ 1 \end{pmatrix}$$

$$\mathbf{v}(1) = \begin{pmatrix} -1 & 0 \\ 0 & -2 \end{pmatrix} \begin{pmatrix} 0 \\ 1 \end{pmatrix} = \begin{pmatrix} 0 \\ -2 \end{pmatrix}$$

$$\mathbf{v}(2) = \begin{pmatrix} 1 & 0 \\ 0 & 4 \end{pmatrix} \begin{pmatrix} 0 \\ 1 \end{pmatrix} = \begin{pmatrix} 0 \\ 4 \end{pmatrix}$$

$$\mathbf{v}(3) = \begin{pmatrix} -1 & 0 \\ 0 & -8 \end{pmatrix} \begin{pmatrix} 0 \\ 1 \end{pmatrix} = \begin{pmatrix} 0 \\ -8 \end{pmatrix}$$

Finally, the closed-form solution is

$$\mathbf{v}(n) = \begin{pmatrix} (-1)^n & 0 \\ 0 & (-2)^n \end{pmatrix} \mathbf{v}(0) \quad n \ge 0$$

Using the z-transform, we have

$$\mathbf{V}(z) = z(z\mathbf{I} - \mathbf{A})^{-1}\mathbf{v}(0) + (z\mathbf{I} - \mathbf{A})^{-1}\mathbf{BX}(z)$$

But again with $\mathbf{x}(n) = 0$, we have

$$(z\mathbf{I} - \mathbf{A}) = \begin{pmatrix} z & 0 \\ 0 & z \end{pmatrix} - \begin{pmatrix} -1 & 0 \\ 0 & -2 \end{pmatrix} = \begin{pmatrix} z+1 & 0 \\ 0 & z+2 \end{pmatrix}$$

$$(z\mathbf{I} - \mathbf{A})^{-1} = \begin{pmatrix} \dfrac{z+2}{(z+1)(z+2)} & \dfrac{0}{(z+1)(z+2)} \\ \dfrac{0}{(z+1)(z+2)} & \dfrac{z+1}{(z+1)(z+2)} \end{pmatrix} = \begin{pmatrix} \dfrac{1}{z+1} & 0 \\ 0 & \dfrac{1}{z+2} \end{pmatrix}$$

Now, we substitute this matrix inversion in the states solution to get

$$\mathbf{V}(z) = z \begin{pmatrix} \dfrac{1}{z+1} & 0 \\ 0 & \dfrac{1}{z+2} \end{pmatrix} \begin{pmatrix} 0 \\ 1 \end{pmatrix} = \begin{pmatrix} 0 \\ \dfrac{z}{z+2} \end{pmatrix}$$

and the inverse of $\mathbf{V}(z)$ is $\mathbf{v}(n)$ which is

$$\mathbf{v}(n) = \begin{pmatrix} 0 \\ (-2)^n u(n) \end{pmatrix}$$

For comparison we find the first few values next.

$$\mathbf{v}(0) = \begin{pmatrix} 0 \\ 1 \end{pmatrix} \quad \mathbf{v}(1) = \begin{pmatrix} 0 \\ -2 \end{pmatrix} \quad \mathbf{v}(2) = \begin{pmatrix} 0 \\ 4 \end{pmatrix} \quad \mathbf{v}(3) = \begin{pmatrix} 0 \\ -8 \end{pmatrix}$$

We can also see that the inverse z-transform of

$$z(z\mathbf{I} - \mathbf{A})^{-1} = \begin{pmatrix} \dfrac{z}{z+1} & 0 \\ 0 & \dfrac{z}{z+2} \end{pmatrix}$$

is

$$\phi(n) = \begin{pmatrix} (-1)^n u(n) & 0 \\ 0 & (-2)^n u(n) \end{pmatrix}$$

as obtained earlier in real time. You can also see here that the transition matrix at $n = 0$ is

$$\phi(0) = \begin{pmatrix} 1 & 0 \\ 0 & 1 \end{pmatrix}$$

which is the identity matrix.

6.7   TRANSFORMATIONS FOR STATE-SPACE REPRESENTATIONS

Let us define another state vector called $\mathbf{w}(n)$ such that

$$\mathbf{w}(n) = \mathbf{P}\mathbf{v}(n) \tag{6.17}$$

where $\mathbf{P}$ is called the transformation matrix. With $\mathbf{w}(n) = \mathbf{P}\mathbf{v}(n)$, we have

$$\mathbf{v}(n) = \mathbf{P}^{-1}\mathbf{w}(n) \tag{6.18}$$

But we know that the original state equation is

$$\mathbf{v}(n+1) = \mathbf{A}\mathbf{v}(n) + \mathbf{B}\mathbf{x}(n)$$

and using the new transformation, we have

$$\mathbf{v}(n+1) = \mathbf{P}^{-1}\mathbf{w}(n+1) \tag{6.19}$$

Then, by substituting in the original state equation, we get

$$\mathbf{P}^{-1}\mathbf{w}(n+1) = \mathbf{A}\mathbf{P}^{-1}\mathbf{w}(n) + \mathbf{B}\mathbf{x}(n) \tag{6.20}$$

Multiply both sides of Equation (5.20) by $\mathbf{P}$ to get

$$\mathbf{w}(n+1) = \mathbf{P}\mathbf{A}\mathbf{P}^{-1}\mathbf{w}(n) + \mathbf{P}\mathbf{B}\mathbf{x}(n) \tag{6.21}$$

and for the output, we have

$$\mathbf{y}(n) = \mathbf{C}\mathbf{P}^{-1}\mathbf{w}(n) + (\mathbf{D})\mathbf{x}(n) \tag{6.22}$$

These are the new state and output equations. The transfer function of the original system is

$$H_{old}(z) = C(zI - A)B + D$$

For the new system, the transfer function is

$$H_{new}(z) = CP^{-1}(zI - PAP^{-1})^{-1}PB + D$$

$$H_{new}(z) = CP^{-1}(zPP^{-1} - PAP^{-1})^{-1}PB + D = CP^{-1}\left[P(zI - A)P^{-1}\right]^{-1}PB + D$$
(6.23)

Knowing that

$$I = PP^{-1} \quad \text{and} \quad (LMN)^{-1} = N^{-1}M^{-1}L^{-1}$$

we write the new transfer function as

$$H_{new}(z) = CP^{-1}P(zI - A)^{-1}P^{-1}PB + D = C(zI - A)^{-1}B + D \qquad (6.24)$$

You can see clearly now that

$$H_{old}(z) = H_{new}(z)$$

This also means that the eigenvalues are the same for the new transformed system.

Assume that the old system has the initial condition vector $v(0)$. Then with $w(n) = Pv(n)$,

$$w(0) = Pv(0) \qquad (6.25)$$

and

$$P^{-1}w(0) = v(0) \qquad (6.26)$$

The solutions of the old system in z-domain for the states and the outputs are

$$V(z) = z(zI - A)^{-1}v(0) + (zI - A)^{-1}BX(z) \qquad (6.27)$$

$$Y(z) = CV(z) + DX(z) \qquad (6.28)$$

and the solutions of the new system in the z-domain are

$$W(z) = z(zI - PAP^{-1})^{-1}w(0) + (zI - PAP^{-1})^{-1}PBX(z) \qquad (6.29)$$

$$Y(z)CP^{-1}W(z) + DX(z) \qquad (6.30)$$

If the matrix $\mathbf{P}$ is the matrix that contains the eigenvectors of $\mathbf{A}$, then $\mathbf{PAP}^{-1}$ is a diagonal or a block diagonal matrix depending on the eigenvalues of $\mathbf{A}$ as we discussed at the beginning of the chapter. When the matrix $\mathbf{A}$ is transformed into a diagonal matrix, where the eigenvalues are located on the main diagonal, the state equations are decoupled and can be solved one by one easily.

6.8 SOME INSIGHTS: POLES AND STABILITY

The objective of this chapter is to represent linear systems in state-space form and to look for ways of solving for the states. In that, the process was to represent an nth-order system (nth-order difference equation) as n first-order difference equations and arrange these equations in what we call state-space representation as

$$\mathbf{v}(n+1) = \mathbf{Av}(n) + \mathbf{BX}(n)$$

$$\mathbf{y}(n) = \mathbf{Cv}(n) + \mathbf{Dx}(n)$$

where

   $\mathbf{x}$ is the input vector (assuming multiple inputs).
   $\mathbf{y}$ is the output vector (assuming multiple outputs).

The $\mathbf{v}$ vector is the vector that contains the states of the system. The $\mathbf{A}$ matrix is the matrix that contains the parameters that control the dynamics of the system. As we saw in previous chapters, in every system representation, there was a way to find the eigenvalues of the system. In state-space representation, the roots of the determinant of the matrix, $(s\mathbf{I} - \mathbf{A})$, where $\mathbf{I}$ is the identity matrix, are the eigenvalues of the system, the poles. And as we mentioned before, these poles determine the shape of the transients of the system under investigation.

   Consider the case where the dynamics matrix $\mathbf{A}$ is

$$\mathbf{A} = \begin{pmatrix} 0 & 1 \\ -6 & -5 \end{pmatrix}$$

$$(\lambda\mathbf{I} - \mathbf{A}) = \begin{pmatrix} \lambda & 0 \\ 0 & \lambda \end{pmatrix} - \begin{pmatrix} 0 & 1 \\ -0 & -5 \end{pmatrix} = \begin{pmatrix} \lambda & -1 \\ 6 & \lambda+5 \end{pmatrix}$$

The eigenvalues are the roots of the determinant of $(\lambda\mathbf{I} - \mathbf{A})$. They are the roots of $\lambda^2 + 5\lambda + 6 = 0$. The eigenvalues are at -3 and -2. Thus, we expect a solution that will contain the terms $c_1(-3)^n + c_2(-2)^n$. The stability of this system depends entirely on the eigenvalues, not on the constants c_1 and c_2. These eigenvalues are the roots of $(\lambda\mathbf{I} - \mathbf{A})$. They are also the roots of the characteristic equation derived from the difference equation and also the roots of the denominator in the transfer function representing the system in the z-domain.

   To summarize, if the system is given in state-space form, the stability of the system is determined by finding the roots of $\det(\lambda\mathbf{I} - \mathbf{A})$. If all the roots are within the unit circle, the system is stable. If one of the roots is not, the system is unstable. And again, the roots will determine the shape of the transients.

6.9 END-OF-CHAPTER EXAMPLES

EOCE 6.1

Consider the following matrices:

$$A = \begin{pmatrix} -2 & 0 \\ 0 & -3 \end{pmatrix} \quad B = \begin{pmatrix} 0 \\ 1 \end{pmatrix} \quad C = \begin{pmatrix} 1 & 2 \end{pmatrix} \quad D = (5)$$

Find the following:

1. $AB - B$ and $(AB)^T + C$
2. A^- and $AA^- 1$
3. The eigenvalues for A and A^2
4. $CDB + D$
5. $BC - A$

SOLUTION

We will use MATLAB® to find the answers to the previous questions. The MATLAB script is EOCE6_1.
The result is

```
ABminusB  =     0
               -4
                    -1   -1
ABtransplusC =       2
invA =  -0.5000      0
             0    -0.3333
           1   0
AinvA =    0   1
          -3
eigA =    -2
          4
eigAA =   9
CDBplusD  =  15
             2   0
BCminusA =   1   5
```

EOCE 6.2

Consider the following matrices:

$$A = \begin{pmatrix} 2 & 9 \\ 0 & 1 \end{pmatrix} \quad B = \begin{pmatrix} 1 & 2 \\ 0 & 1 \end{pmatrix} \quad C = \begin{pmatrix} 1 & 0 \\ 0 & 1 \end{pmatrix}$$

Find

1. Eigenvalues for **A** and **B**
2. **A⁻** and **B⁻¹**
3. **(AB)⁻** and **B⁻ A⁻¹**
4. **BCBC** and **CA⁻¹**
5. **(CA)⁻¹** and **(BC)⁻¹ A**

SOLUTION

MATLAB is used again here. The script is EOCE6_2.
The result is

```
Ainv =    0.5000   -4.5000
          0          1.0000

          1   -2
Binv =
          0    1

          0.5000   -6.5000
ABinv =
          0          1.0000

BinvAinv =    0.5000   -6.5000
              0          1.0000

          1   4
BCBC =
          0   1

          0.5000   -4.5000
CinvA =
          0          1.0000

CAinv =   0.5000   -4.5000
          0          1.0000

          2   7
BCinvA =
          0   1
```

EOCE 6.3

Consider the following matrices:

$$A = \begin{pmatrix} -1 & 0 \\ 0 & -2 \end{pmatrix} \quad B = \begin{pmatrix} -3 & 0 \\ 0 & 4 \end{pmatrix} \quad C = \begin{pmatrix} 1 & 1 \\ 0 & 2 \end{pmatrix}$$

1. Find eigenvalues and eigenvectors for **A** and **B**.
2. Find eigenvalues and eigenvectors for **A²** and **B²**.
3. Put **C** in the diagonal form by first calculating **P**, the matrix of the eigenvector for **C**.
4. What are the eigenvectors and eigenvalues of **P⁻¹CP**?

SOLUTION

We will use MATLAB again. The MATLAB command

```
[V, D] = eig(A, B)
```

produces a diagonal matrix **D** of eigenvalues usually on the main diagonal and a matrix **V** with eigenvector columns. The script is EOCE6_3.

The result is

$$Aeigvectors = \begin{matrix} 0 & 1 \\ 1 & 0 \end{matrix}$$

$$Aeigenvalues = \begin{matrix} -2 & 0 \\ 0 & -1 \end{matrix}$$

$$Beigvectors = \begin{matrix} 1 & 0 \\ 0 & 1 \end{matrix}$$

$$Beigenvalues = \begin{matrix} -3 & 0 \\ 0 & 4 \end{matrix}$$

$$AAeigvectors = \begin{matrix} 1 & 0 \\ 0 & 1 \end{matrix}$$

$$AAeigenvalues = \begin{matrix} 1 & 0 \\ 0 & 4 \end{matrix}$$

$$BBeigvectors = \begin{matrix} 1 & 0 \\ 0 & 1 \end{matrix}$$

$$BBeigenvalues = \begin{matrix} 9 & 0 \\ 0 & 16 \end{matrix}$$

$$Pofeigenvectors = \begin{matrix} 1.0000 & 07071 \\ 0 & 0.7071 \end{matrix}$$

$$Cindiagonalform = \begin{matrix} 1 & 0 \\ 0 & 2 \end{matrix}$$

$$PinvCPeigvectors = \begin{matrix} 1 & 0 \\ 0 & 1 \end{matrix}$$

$$PinvCPeigenvalues = \begin{matrix} 1 & 0 \\ 0 & 2 \end{matrix}$$

EOCE 6.4

Put the following system in state-space form:

$$y(n) - 2y(n-1) + 3y(n-3) = 4x(n)$$

SOLUTION

Let

$$v_1(n) = y(n-3)$$

$$v_2(n) = y(n-2)$$

$$v_3(n) = y(n-1)$$

We have three states since our system is third order. Thus, we have

$$v_1(n+1) = y(n-2) = v_2(n)$$

$$v_2(n+1) = y(n-1) = v_3(n)$$

$$v_3(n+1) = 2y(n-1) - 3y(n-3) + 4x(n) = 2v_3(n) - 3v_1(n) + 4x(n)$$

The output $y(n)$ is

$$y(n) = 2v_3(n) - 3v_1(n) + 4x(n)$$

and the state and output equations are

$$\mathbf{v}(n+1) = \begin{pmatrix} 0 & 1 & 0 \\ 0 & 0 & 1 \\ -3 & 0 & 2 \end{pmatrix} \mathbf{v}(n) + \begin{pmatrix} 0 \\ 0 \\ 4 \end{pmatrix} \mathbf{x}(n)$$

$$\mathbf{y}(n) = \begin{pmatrix} -3 & 0 & 2 \end{pmatrix} \mathbf{v}(n) + (4)\mathbf{x}(n)$$

EOCE 6.5

Consider the following system:

$$H(z) = \frac{2z+1}{z^3 + 3z^2 + 2z + 8}$$

Write the state and output equations for this system.

SOLUTION

We can deduce the difference equation from $H(z)$ first, then find the state equations. From the transfer function, we can write

$$Y(z)\left[z^3 + 3z^2 + 2z + 1\right] = \left[2z + 1\right]X(z)$$

By taking the inverse transform, we get

$$y(n) + 2y(n+1) + 3y(n+2) + y(n+3) = x(n) + 2x(n+1)$$

Let us draw the block diagram first. The block diagram is shown in Figure 6.8. From the figure, we see that the states are

$$v_3(n+1) = x(n) - 3v_3(n) - 2v_2(n) - v_1(n)$$

$$v_2(n+1) = v_3(n)$$

$$v_1(n+1) = v_2(n)$$

and the output is

$$y(n) = v_1(n) + 2v_2(n)$$

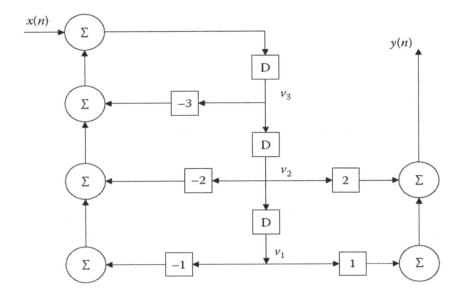

FIGURE 6.8 Block diagram for EOCE 6.5.

From the state and the output equations, we can form the state-space matrix equations as

$$\mathbf{v}(n+1) = \begin{pmatrix} 0 & 1 & 0 \\ 0 & 0 & 1 \\ -1 & -2 & -3 \end{pmatrix} \mathbf{v}(n) + \begin{pmatrix} 0 \\ 0 \\ 1 \end{pmatrix} \mathbf{x}(n)$$

$$\mathbf{y}(n) = \begin{pmatrix} 1 & 2 & 0 \end{pmatrix} \mathbf{v}(n) + (0)\mathbf{x}(n)$$

We can also write different state equations using Example 6.2. Referring to Example 6.2, we have

$$H(z) = \frac{b_0 z^3 + b_1 z^2 + b_2 z + b_3}{z^3 + a_1 z^2 + a_2 z + a_3} = \frac{0z^3 + 0z^2 + 2z + 1}{z^3 + 3z^2 + 2z + 1}$$

in our present case. Then by inspection, we have

$$\mathbf{v}(n+1) = \begin{pmatrix} 0 & 1 & 0 \\ 0 & 0 & 1 \\ -1 & -2 & -3 \end{pmatrix} \mathbf{v}(n) + \begin{pmatrix} 0 \\ 0 \\ 1 \end{pmatrix} \mathbf{x}(n)$$

$$\mathbf{y}(n) = \begin{pmatrix} (b_3 - a_3(b_0)) & b_2 - a_2(b_0) & b_1 - a_1(b_0) \end{pmatrix} \mathbf{v}(n) + b_0 \mathbf{x}(n)$$

After substitution, we arrive at

$$y(n) = \begin{pmatrix} 1 & 2 & 0 \end{pmatrix} v(n) + (0) x(n)$$

EOCE 6.6

Consider the following system in Figure 6.9. Write the state equations in matrix form.

SOLUTION

Let the output of the first delay be $v_1(n)$, $v_2(n)$ for the second delay, and $v_3(n)$ for the third delay. Then the state equations are

$$v_1(n++1) = x(n) - 2v_1(n)$$

$$v_2(n+1) = x(n) - 3v_2(n) + v_1(n)$$

$$v_3(n+1) = x(n) - 5v_3(n) + v_2(n)$$

The output equation is

$$y(n) = x(n) + v_3(n)$$

The state and output equations in state-space matrix form then are

$$v(n+1) = \begin{pmatrix} -2 & 0 & 0 \\ 1 & -3 & 0 \\ 0 & 0 & -5 \end{pmatrix} v(n) + \begin{pmatrix} 1 \\ 1 \\ 1 \end{pmatrix} x(n)$$

$$y(n) = \begin{pmatrix} 0 & 0 & 1 \end{pmatrix} v(n) + x(n)$$

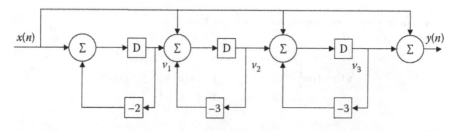

FIGURE 6.9   Block diagram for EOCE 6.6.

EOCE 6.7

Consider the following system:

$$\mathbf{v}(n+1) = \begin{pmatrix} 1 & 0 \\ 1 & k \end{pmatrix} \mathbf{v}(n) + \begin{pmatrix} 0 \\ 1 \end{pmatrix} x(n)$$

$$\mathbf{y}(n) = \begin{pmatrix} 0 & 1 \end{pmatrix} \mathbf{v}(n)$$

1. For what values of k is the system stable?
2. For a value of k that stabilizes the system, find $H(z) = Y(z)/X(z)$.
3. Use recursion to solve for $\mathbf{v}(0)$, $\mathbf{v}(1)$, $\mathbf{v}(2)$, $\mathbf{v}(3)$, and $\mathbf{v}(4)$ with $x(n) = u(n)$ and $\mathbf{v}(0) = \begin{pmatrix} 0 \\ 1 \end{pmatrix}$.
4. Use the z-transform to solve for $\mathbf{v}(n)$ for $n \geq 0$.
5. What is the transition matrix?

SOLUTION

1. The eigenvalues of this system are the roots of the characteristic equation, which is the determinant of $(\lambda\mathbf{I} - \mathbf{A})$ set equal to zero.

$$(\lambda\mathbf{I} - \mathbf{A}) = \begin{pmatrix} \lambda & 0 \\ 0 & \lambda \end{pmatrix} - \begin{pmatrix} 1 & 0 \\ 1 & k \end{pmatrix} = \begin{pmatrix} \lambda-1 & 0 \\ -1 & \lambda-k \end{pmatrix}$$

$$\det(\lambda\mathbf{I} - \mathbf{A}) = (\lambda - 1)(\lambda - k)$$

This system is stable if the eigenvalues of the system matrix $\mathbf{A}$ are all within the unit circle. The eigenvalues are at 1 and k. Therefore, for stability, the value of k has to be within the unit circle.

2. For $k = -1/2$, the state matrices are

$$\mathbf{A} = \begin{pmatrix} 1 & 0 \\ 1 & -\dfrac{1}{2} \end{pmatrix} \quad \mathbf{B} = \begin{pmatrix} 0 \\ 1 \end{pmatrix} \quad \mathbf{C} = \begin{pmatrix} 0 & 1 \end{pmatrix} \quad \mathbf{D} = (0)$$

The transfer function in this case is

$$H(z) = \mathbf{C}(z\mathbf{I} - \mathbf{A})^{-1}\mathbf{B} + \mathbf{D}$$

$$H(z) = \begin{pmatrix} 0 & 1 \end{pmatrix} \begin{pmatrix} z-1 & 0 \\ -1 & z+\dfrac{1}{2} \end{pmatrix}^{-1} \begin{pmatrix} 0 \\ 1 \end{pmatrix}$$

$$H(z) = \begin{pmatrix} 0 & 1 \end{pmatrix} \dfrac{1}{(z+1/2)(z-1)} \begin{pmatrix} z+\dfrac{1}{2} & 0 \\ 1 & z-1 \end{pmatrix} \begin{pmatrix} 0 \\ 1 \end{pmatrix}$$

$$H(z) = \begin{pmatrix} \dfrac{1}{(z+1/2)(z-1)} & \dfrac{z-1}{(z+1/2)(z-1)} \end{pmatrix} \begin{pmatrix} 0 \\ 1 \end{pmatrix} = \dfrac{z-1}{(z+1/2)(z-1)}$$

We use MATLAB to find the transfer function **H**(z) from the matrices **A**, **B**, **C**, and **D**. We will use the MATLAB function ss2tf (state space to transfer function) to do that in the script EOCE6_71.
    to get

```
num = 0   1   -1

den = 1.0000   -0.5000   -0.5000
```

and the transfer function is

$$H(z) = \dfrac{z-1}{\left(z^2 - (1/2)z - 1/2\right)} = \dfrac{z-1}{(z+1/2)(z-1)}$$

3. To use recursion to solve for **v**(n), we will use the equation

$$\mathbf{v}(n+1) = \mathbf{A}\mathbf{v}(n) + \mathbf{B}\mathbf{x}(n)$$

and

$$\mathbf{y}(n) = \mathbf{C}\mathbf{v}(n) + \mathbf{D}\mathbf{x}(n)$$

We will have the initial sate as

$$\mathbf{v}(0) = \begin{pmatrix} 0 \\ 1 \end{pmatrix}$$

$$\mathbf{y}(0) = \begin{pmatrix} 0 & 1 \end{pmatrix} \begin{pmatrix} 0 \\ 1 \end{pmatrix} = 1$$

The other few terms of the state vector and the output are given as follows:

$$\mathbf{v}(1) = \begin{pmatrix} 1 & 0 \\ 1 & -\dfrac{1}{2} \end{pmatrix}\begin{pmatrix} 0 \\ 1 \end{pmatrix} + \begin{pmatrix} 0 \\ 1 \end{pmatrix}(1) = \begin{pmatrix} 0 \\ -\dfrac{1}{2} \end{pmatrix} + \begin{pmatrix} 0 \\ 1 \end{pmatrix} = \begin{pmatrix} 0 \\ \dfrac{1}{2} \end{pmatrix}$$

$$\mathbf{y}(1) = \begin{pmatrix} 0 & 1 \end{pmatrix}\begin{pmatrix} 0 \\ 1 \\ 2 \end{pmatrix} = \dfrac{1}{2}$$

$$\mathbf{v}(2) = \begin{pmatrix} 0 & 0 \\ 1 & -\dfrac{1}{2} \end{pmatrix}\begin{pmatrix} 0 \\ \dfrac{1}{2} \end{pmatrix} + \begin{pmatrix} 0 \\ 1 \end{pmatrix}(1) = \begin{pmatrix} 0 \\ -\dfrac{1}{4} \end{pmatrix} + \begin{pmatrix} 0 \\ 1 \end{pmatrix} = \begin{pmatrix} 0 \\ \dfrac{3}{4} \end{pmatrix}$$

$$\mathbf{y}(2) = \begin{pmatrix} 0 & 1 \end{pmatrix}\begin{pmatrix} 0 \\ \dfrac{3}{4} \end{pmatrix} = \dfrac{3}{4}$$

and so on. We can use MATLAB to do this recursion. We will do that in the script EOCE6_72 to get the same results as we had before.
4. We can also use the z-transform method to find the output and the state values

$$\mathbf{V}(z) = z(z\mathbf{I} - \mathbf{A})^{-1}\mathbf{v}(0) + (z\mathbf{I} - \mathbf{A})^{-1}\mathbf{B}X(z)$$

$$\mathbf{V}(z) = \dfrac{z}{(z+1/2)(z-1)}\begin{pmatrix} z+\dfrac{1}{2} & 0 \\ 1 & z-1 \end{pmatrix}\begin{pmatrix} 0 \\ 1 \end{pmatrix}$$

$$+5\dfrac{1}{(z+1/2)(z-1)}\begin{pmatrix} z+\dfrac{1}{2} & 0 \\ 1 & z-1 \end{pmatrix}\begin{pmatrix} 0 \\ 1 \end{pmatrix}\dfrac{z}{z-1}$$

$$\mathbf{V}(z) = \begin{pmatrix} 0 \\ \dfrac{z(z-1)}{(z+1/2)(z-1)} \end{pmatrix} + \begin{pmatrix} 0 \\ \dfrac{z-1}{(z+1/2)(z-1)} \end{pmatrix}\dfrac{z}{z-1}$$

After some simplifications, we arrive at

$$\mathbf{V}(z) = \begin{pmatrix} 0 \\ \dfrac{z(z-1)+z}{(z+1/2)(z-1)} \end{pmatrix} = \begin{pmatrix} 0 \\ \dfrac{z^2}{(z+1/2)(z-1)} \end{pmatrix}$$

and the output in the z-domain is given by

$$\mathbf{Y}(z) = \mathbf{CV}(z) = \begin{pmatrix} 0 & 1 \end{pmatrix} v_z = \frac{z^2}{(z+1/2)(z-1)}$$

The output is one-dimensional or scalar. We can do partial fraction expansion using MATLAB and write the script EOCE6_73. to get

```
r = 0.6667   0.3333
p = 1.0000  -0.500
```

Therefore, with the help of MATLAB, we arrive at

$$Y(z) = \frac{0.6667z}{z-1} + \frac{z(1/3)}{z+(1/2)}$$

and by taking the inverse transform, we get

$$y(n) = v_2(n) = \left[0.6667(1)^n + \frac{1}{3}\left(-\frac{1}{2}\right)^n \right] u(n)$$

If we substitute values for n, we can verify the results we arrived at earlier. Try that.

5. The transition matrix $\mathbf{A}^n$ is the inverse transform of

$$z(z\mathbf{I} - \mathbf{A})^{-1} = \begin{pmatrix} \dfrac{z}{(z-1)} & 0 \\[3mm] \dfrac{z}{(z+1/2)(z-1)} & \dfrac{z}{z+1/2} \end{pmatrix}$$

$$z(z\mathbf{I} - \mathbf{A})^{-1} = \begin{pmatrix} \dfrac{z}{(z-1)} & 0 \\[3mm] \dfrac{0.6667}{(z-1)} + \dfrac{1/3}{z+1/2} & \dfrac{z}{(z+1/2)} \end{pmatrix}$$

By taking the inverse z-transform, we find the transition matrix as

$$\mathbf{A}^n = \begin{pmatrix} (1)^n & 0 \\[3mm] 0.6667(1)^n + \left(\dfrac{1}{3}\right)\left(-\dfrac{1}{2}\right)^n & \left(-\dfrac{1}{2}\right)^n \end{pmatrix} u(n)$$

You can see that $\mathbf{A}^n = \begin{pmatrix} 1 & 0 \\ 0 & 1 \end{pmatrix}$ as an indication that the transition matrix is correct.

We can also use MATLAB to verify the entries in the transition matrix $\mathbf{A}^n$. For the first entry, we have $\varphi_{11}(z) = z/(z-1)$. We will write the MATLAB script EOCE6_74 to find $\varphi_{11}(n)$.

The result is $1u(n)$ as expected.

EOCE 6.8

Consider the system in Figure 6.10.

1. Write down the state and output equations.
2. What is the transition matrix?
3. With $\mathbf{x}(n) = 0$ and an initial state vector of $\begin{pmatrix} 1 \\ 0 \end{pmatrix}$, what is the state vector and the output?
4. Is the system stable?
5. Use MATLAB to find $\mathbf{y}(n)$, for $\mathbf{x}(n) = u(n)$ and the initial state of $\begin{pmatrix} 1 \\ 0 \end{pmatrix}$.
6. Find the transfer function $\mathbf{H}(z)$.

SOLUTION

1. From Figure 6.10 and by taking $v_1(n)$ as the output of the upper delay and $v_2(n)$ as the output of the lower delay, we have the state equations

$$v_1(n+1) = x(n) - v_1(n)$$

$$v_2(n+1) = x(n) - 2v_2(n)$$

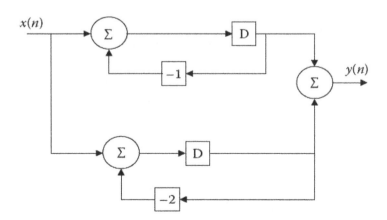

FIGURE 6.10 Block diagram for EOCE 6.8.

The output equation is

$$y(n) = v_1(n) + v_2(n)$$

Then the state and output equations are grouped as

$$v(n+1) = \begin{pmatrix} -1 & 0 \\ 0 & -2 \end{pmatrix} v(n) \begin{pmatrix} 1 \\ 1 \end{pmatrix} x(n)$$

$$y(n) = \begin{pmatrix} 1 & 1 \end{pmatrix} v(n) + (0) x(n)$$

2. With $A = \begin{pmatrix} -1 & 0 \\ 0 & -2 \end{pmatrix}$, the state transition matrix is the inverse trans-

form of $z(zI - A)^{-1}$. We have

$$(zI - A)^{-1} = \begin{pmatrix} z+1 & 0 \\ 0 & z+2 \end{pmatrix}^{-1} = \begin{pmatrix} z+2 & 0 \\ 0 & z+1 \end{pmatrix} \frac{1}{(z+1)(z+2)}$$

and

$$(zI - A)^{-1} = \begin{pmatrix} \dfrac{z}{z+1} & 0 \\ 0 & \dfrac{z}{z+2} \end{pmatrix}$$

The transition matrix written is the inverse transform of $z(zI - A)^{-1}$ and is

$$A^n = \begin{pmatrix} (-1)^n u(n) & 0 \\ 0 & (-2)^n u(n) \end{pmatrix}$$

But since A is in diagonal form, we could have found A^n by inspection. With

$$A = \begin{pmatrix} -1 & 0 \\ 0 & -2 \end{pmatrix} \qquad A^n = \begin{pmatrix} (-1)^n u(n) & 0 \\ 0 & (-2)^n u(n) \end{pmatrix}$$

3. With $x(n) = 0$, the state vector becomes

$$v(n+1) = Av(n)$$

For $n = 0$, we have

$$\mathbf{v}(0) = \begin{pmatrix} 1 \\ 0 \end{pmatrix}$$

$$\mathbf{y}(0) = v_1(0) + v_2(0) = 1$$

For $n = 1$, we get

$$\mathbf{v}(1) = \begin{pmatrix} -1 & 0 \\ 0 & -2 \end{pmatrix}\begin{pmatrix} 1 \\ 0 \end{pmatrix} = \begin{pmatrix} -1 \\ 0 \end{pmatrix}$$

$$\mathbf{y}(1) = v_1(1) + v_2(1) = -1 + 0 = -1$$

For $n = 2$, we have

$$\mathbf{v}(2) = \begin{pmatrix} -1 & 0 \\ 0 & -2 \end{pmatrix}\begin{pmatrix} 1 \\ 0 \end{pmatrix} = \begin{pmatrix} 1 \\ 0 \end{pmatrix}$$

$$\mathbf{y}(2) = v_1(2) + v_2(2) = 1 + 0 = 1$$

For $n = 3$, we get

$$\mathbf{v}(3) = \begin{pmatrix} -1 & 0 \\ 0 & -2 \end{pmatrix}\begin{pmatrix} 1 \\ 0 \end{pmatrix} = \begin{pmatrix} -1 \\ 0 \end{pmatrix}$$

$$\mathbf{y}(3) = v_1(3) + v_2(3) = 1 + 0 = -1$$

For $n = 4$, we get

$$\mathbf{v}(4) = \begin{pmatrix} -1 & 0 \\ 0 & -2 \end{pmatrix}\begin{pmatrix} -1 \\ 0 \end{pmatrix} = \begin{pmatrix} 1 \\ 0 \end{pmatrix}$$

$$\mathbf{y}(4) = v_1(4) + v_2(4) = 1 + 0 = 1$$

Then by induction, we can see that

$$\mathbf{v}(n) = \begin{pmatrix} (-1)^n u(n) \\ 0 \end{pmatrix}$$

and

$$\mathbf{y}(n) = (-1)^n u(n)$$

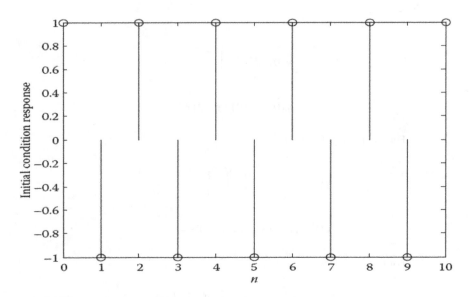

FIGURE 6.11   Plot for EOCE 6.8.

Using MATLAB, we can use recursion and write the script EOCE6_81 to solve for the output and the state vector.

And the result will be identical to what we just found. We can also plot y(n) versus n by using the MATLAB function dlsim as in the EOCE6_82 script.

The plots are shown in Figure 6.11.

4. The stability of the system depends on the location of the poles. One of the poles is not within the unit circle. Thus, the system is not stable.
5. With $x(n) = u(n)$ and the initial conditions, $v_1(0) = 1$ and $v_2(0) = 0$, we can use MATLAB to find $y(n)$ as in the EOCE6_83 script. The plots are shown in Figure 6.12.
6. The transfer function $\mathbf{H}(z)$ is calculated using the equation

$$\mathbf{H}(z) = \mathbf{C}(z\mathbf{I} - \mathbf{A})^{-1}\mathbf{B} + \mathbf{D}$$

$$\mathbf{H}(z) = \begin{pmatrix} 1 & 1 \end{pmatrix} \begin{pmatrix} \dfrac{1}{z+1} & 0 \\[2mm] 0 & \dfrac{1}{z+2} \end{pmatrix} \begin{pmatrix} 1 \\ 1 \end{pmatrix} + (0) = \begin{pmatrix} \dfrac{1}{z+1} & \dfrac{1}{z+2} \end{pmatrix} \begin{pmatrix} 1 \\ 1 \end{pmatrix}$$

Finally,

$$\mathbf{H}(z) = \frac{1}{z+1} + \frac{1}{z+2} = \frac{z+z+2+1}{(z+1)(z+2)} = \frac{2z+3}{z^2+3z+2}$$

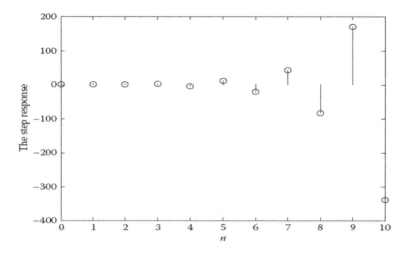

FIGURE 6.12 Plot for EOCE 6.8.

We can use MATLAB to verify this result as in the EOCE6_84 script.
to get
num = 0 2 3
den = 1 3 2
which verifies the result.

EOCE 6.9

We have seen in this chapter that the state-space representation is not unique. To illustrate that, consider the following system

$$v(n+1) = Av(n) + Bx(n)$$

$$y(n) = Cv(n) + (D)x(n)$$

With

$$A = \begin{pmatrix} 2 & 0 \\ 0 & 3 \end{pmatrix} \quad B = \begin{pmatrix} 0 \\ 1 \end{pmatrix} \quad C = \begin{pmatrix} 0 & 1 \end{pmatrix} \quad D = (0)$$

What are the other state-space representations?

SOLUTION

Let us define another state vector called $w(n)$ such that $w(n) = Pv(n)$, where P is called the transformation matrix. Let

$$P = \begin{pmatrix} 1 & 0 \\ 0 & 2 \end{pmatrix}$$

Then the new system is

$$\mathbf{w}(n+1) = \mathbf{PAP}^{-1}\mathbf{w}(n) + \mathbf{PBx}(n)$$

and for the output, we have

$$\mathbf{y}(n) = \mathbf{CP}^{-1}\mathbf{w}(n) + (\mathbf{D})\mathbf{x}(n)$$

Assume that the old system has the initial condition vector $v(0) = \begin{pmatrix} 1 \\ 0 \end{pmatrix}$. Then with $\mathbf{w}(n) = \mathbf{Pv}(n)$, $\mathbf{w}(0) = \mathbf{Pv}(0)$, and $\mathbf{P}^{-1}\mathbf{w}(0) = \mathbf{v}(0)$. The solutions of the new system in the z-domain are

$$\mathbf{W}(z) = z(z\mathbf{I} - \mathbf{PAP}^{-1})^{-1}\mathbf{w}(0) + (z\mathbf{I} - \mathbf{PAP}^{-1})^{-1}\mathbf{PBX}(z)$$

and

$$\mathbf{Y}(z) = \mathbf{CP}^{-1}\mathbf{W}(z) + \mathbf{DX}(z)$$

Let us now use MATLAB to find the step response with zero initial conditions to the old system; we will then consider the new system and see that the two outputs are the same. The MATLAB script is EOCE6_91.

The plots are shown in Figure 6.13.

We can also use MATLAB to check the stability and the eigenvalues for both systems as in the EOCE6_92 script.

to get

```
Aeigenvalues =    0.4000
                  0.5000

Aneweigenvalues = 0.4000
                  0.5000
```

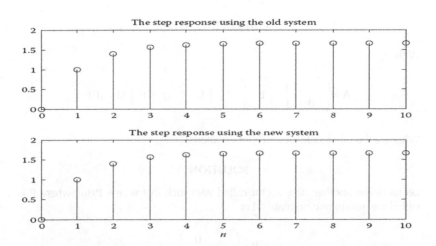

FIGURE 6.13 Plot for EOCE 6.9.

EOCE 6.10

Consider the following system:

$$y(n) + y(n-1) + y(n-2) = x_1(n) + x_2(n)$$

1. Is the system stable?
2. Find $y(n)$ if $x_1(n) = (.1)^n u(n)$ and $x_2(n) = (.1)^n u(n)$.
3. Find the transfer function $H(z)$.

SOLUTION

1. Let us write the state and output equations first. Let

$$v_1(n) = y(n-2)$$
$$v_2(n) = y(n-1)$$

Then,

$$v_1(n+1) = y(n-1) = v_2(n)$$

and

$$v_2(n+1) = y(n) = x_1(n) + x_2(n) - y(n-1) - y(n-2)$$
$$= x_1(n) + x_2(n) - v_2(n) - v_1(n)$$

The state and output equations are then

$$v(n+1) = \begin{pmatrix} 0 & 0 \\ -1 & -1 \end{pmatrix} v(n) + \begin{pmatrix} 0 & 0 \\ 1 & 1 \end{pmatrix} \begin{pmatrix} x_1(n) \\ x_2(n) \end{pmatrix}$$

$$y(n) = \begin{pmatrix} -1 & -1 \end{pmatrix} v(n) + \begin{pmatrix} 1 & 1 \end{pmatrix} \begin{pmatrix} x_1(n) \\ x_2(n) \end{pmatrix}$$

with

$$A = \begin{pmatrix} 0 & 1 \\ -1 & -1 \end{pmatrix} \quad B = \begin{pmatrix} 0 & 0 \\ 1 & 1 \end{pmatrix} \quad C = \begin{pmatrix} -1 & -1 \end{pmatrix} \quad D = \begin{pmatrix} 1 & 1 \end{pmatrix}$$

The eigenvalues for A are the roots of the determinant of $(z\mathbf{I} - A)$.

$$\det(z\mathbf{I} - A) = \det\left(\begin{pmatrix} z & -1 \\ 1 & z+1 \end{pmatrix}\right)$$

$$\det(z\mathbf{I} - A) = z^2 + z + 1$$

The roots are

$$z_{1,2} = \frac{-1 \pm \sqrt{1-4}}{2} = \frac{-1}{2} \pm j\sqrt{\frac{3}{2}}$$

The magnitude of the roots is unity. This means that the system is on the verge of instability or we can call it unstable.

2. With zero initial conditions,

$$\mathbf{V}(z) = (z\mathbf{I} - \mathbf{A})^{-1} \mathbf{B} \mathbf{X}(z)$$

$$\mathbf{Y}(z) = \mathbf{C}\mathbf{V}(z) + \mathbf{D}\mathbf{X}(z)$$

We can substitute in the previous state equation and write

$$\mathbf{V}(z) = \begin{pmatrix} \dfrac{z+1}{z^2+z+1} & \dfrac{1}{z^2+z+1} \\ \dfrac{-1}{z^2+z+1} & \dfrac{z}{z^2+z+1} \end{pmatrix} \begin{pmatrix} 0 & 0 \\ 1 & 1 \end{pmatrix} \begin{pmatrix} \dfrac{z}{z-.1} \\ \dfrac{z}{z+.1} \end{pmatrix}$$

$$= \begin{pmatrix} \dfrac{1}{z^2+z+1} & \dfrac{1}{z^2+z+1} \\ \dfrac{z}{z^2+z+1} & \dfrac{z}{z^2+z+1} \end{pmatrix} \begin{pmatrix} \dfrac{z}{z-.1} \\ \dfrac{z}{z+.1} \end{pmatrix}$$

$$\mathbf{V}(z) = \begin{pmatrix} \dfrac{z}{(z-.1)(z^2+z+1)} + \dfrac{z}{(z+.1)(z^2+z+1)} \\ \dfrac{z^2}{(z-.1)(z^2+z+1)} + \dfrac{z}{(z+.1)(z^2+z+1)} \end{pmatrix}$$

The output equation in the z-domain is

$$\mathbf{Y}(z) = \begin{pmatrix} -1 & -1 \end{pmatrix} \mathbf{V}(z) + \begin{pmatrix} 1 & 1 \end{pmatrix} \begin{pmatrix} \dfrac{z}{z-.1} \\ \dfrac{z}{z+.1} \end{pmatrix}$$

or

$$\mathbf{Y}(z) = \frac{-z}{(z-.1)(z^2+z+1)} + \frac{-z}{(z+.1)(z^2+z+1)} + \frac{-z^2}{(z-.1)(z^2+z+1)}$$

$$+ \frac{-z}{(z+.1)(z^2+z+1)} + \frac{z}{z-.1} + \frac{z}{z+.1}$$

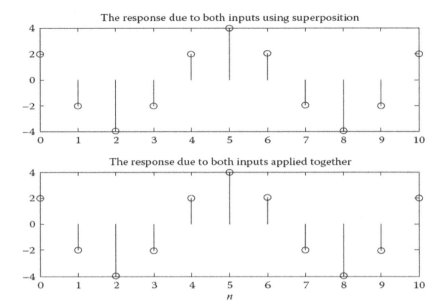

FIGURE 6.14 Plots for EOCE 6.10.

We can use MATLAB to find the inverse transform of $\mathbf{Y}(z)$ and get $\mathbf{y}(n)$. We will leave that as an exercise.

We can also use MATLAB to find $y(n)$ as in the following script. We will use superposition first. This means that we will kill the inputs one at a time. In this case, only the $\mathbf{B}$ and the $\mathbf{D}$ matrices will change.

$$\mathbf{A} = \begin{pmatrix} 0 & 1 \\ -1 & -1 \end{pmatrix} \mathbf{B} = \begin{pmatrix} 0 \\ 1 \end{pmatrix} \mathbf{C} = \begin{pmatrix} -1 & -1 \end{pmatrix} \mathbf{D} = (1)$$

We will then apply both inputs at once. The matrices in this case are

$$\mathbf{A} = \begin{pmatrix} 0 & 1 \\ -1 & -1 \end{pmatrix} \mathbf{B} = \begin{pmatrix} 0 & 0 \\ 1 & 1 \end{pmatrix} \mathbf{C} = \begin{pmatrix} -1 & -1 \end{pmatrix} \mathbf{D} = \begin{pmatrix} 1 & 1 \end{pmatrix}$$

The MATLAB script is EOCE6_101.
The plots are shown in Figure 6.14.
3. The transfer function $\mathbf{H}(z)$ is

$$\mathbf{H}(z) = \mathbf{C}(z\mathbf{I} - \mathbf{A})^{-1}\mathbf{B} + \mathbf{D}$$

$$= \begin{pmatrix} -1 & -1 \end{pmatrix} \begin{pmatrix} \dfrac{z+1}{z^2+z+1} & \dfrac{1}{z^2+z+1} \\ \dfrac{-1}{z^2+z+1} & \dfrac{z}{z^2+z+1} \end{pmatrix} \begin{pmatrix} 0 & 0 \\ 1 & 1 \end{pmatrix} + \begin{pmatrix} 1 & 1 \end{pmatrix}$$

$$H(z) = \left(-\frac{-(z+1)}{z^2+z+1} + \frac{1}{z^2+z+1} \quad \frac{-1}{z^2+z+1} + \frac{-z}{z^2+z+1} \right) \left(\begin{array}{cc} 0 & 0 \\ 1 & 1 \end{array} \right) + \left(\begin{array}{cc} 1 & 1 \end{array} \right)$$

$$H(z) = \left(\frac{-1}{z^2+z+1} + \frac{-z}{z^2+z+1} \quad \frac{-1}{z^2+z+1} + \frac{-z}{z^2+z+1} \right) + \left(\begin{array}{cc} 1 & 1 \end{array} \right)$$

Since we had two inputs and single output, we should have one row in $H(z)$ with two entries. They are

$$H_{11}(z) = \frac{-1}{z^2+z+1} + \frac{(-z)}{z^2+z+1} + 1 = \frac{z^2}{z^2+z+1} = \frac{Y(z)}{X_1(z)}$$

$$H_{12}(z) = \frac{-1}{z^2+z+1} + \frac{(-z)}{z^2+z+1} + 1 = \frac{z^2}{z^2+z+1} = \frac{Y(z)}{X_2(z)}$$

We can also use MATLAB to find $H(z)$ as in the script EOCE6_102. The result is

```
num1 =1  0  0
den  = 1.0000 1.0000 1.0000
num2 = 1 0 0
den  = 1.0000 1.0000 1.0000
```

which agrees with the analytical results obtained previously.

EOCE 6.11

Consider the following system:

$$H(z) = \frac{z^3 + 3z^2 + 2z + 0}{z^3 - 6z^2 + 11z - 6}$$

Find the output $y(n)$ if the input is $x(n) = u(n)$. Use as many different state-space representations as you wish.

SOLUTION

1. Using Example 6.2 we can write the first state-space representation by inspection. We have the state equation as

$$v(n+1) = \left(\begin{array}{ccc} 0 & 1 & 0 \\ 0 & 0 & 1 \\ 6 & -11 & 6 \end{array} \right) v(n) + \left(\begin{array}{c} 0 \\ 0 \\ 1 \end{array} \right) x(n)$$

and the output equation as

$$y(n) = \left[0 - (-6)(1) \ 2 - 11(1) \ 3 - (-6)(1)\right] v(n) + (1)x(n)$$

$$= \left(\begin{array}{ccc} 6 & -9 & -9 \end{array}\right) v(n) + (1)x(n)$$

2. Series connections. The transfer function can be written as

$$H(z) = \frac{z(z^2 + 3z + 1)}{(z-2)(z-3)(z-1)} = \frac{z(z+2)(z+1)}{(z-2)(z-3)(z-1)}$$

$$H(z) = \frac{z}{z-2} \frac{z+2}{z-3} \frac{z+1}{z-1}$$

The block diagram is shown in Figure 6.15. From Figure 6.15, we have the state equations as

$$v_3(n+1) = x(n) + 2v_3(n)$$

$$v_2(n+1) = x(n) + 2v_3(n) + 3v_2(n)$$

$$v_1(n+1) = 2v_2(n) + x(n) + 2v_3(n) + 3v_2(n) + v_1(n)$$

and the output equation is

$$y(n) = 2v_2(n) + x(n) + 2v_3(n) + 3v_2(n) + 2v_1(n)$$

Thus, the state-space system is

$$v(n+1) = \left(\begin{array}{ccc} 1 & 5 & 2 \\ 0 & 3 & 2 \\ 0 & 0 & 2 \end{array}\right) v(n) + \left(\begin{array}{c} 1 \\ 1 \\ 1 \end{array}\right) x(n)$$

$$y(n) = \left(\begin{array}{ccc} 2 & 5 & 2 \end{array}\right) v(n) + (1)x(n)$$

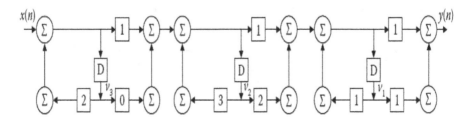

FIGURE 6.15 Block diagram for EOCE 6.11.

3. Using partial fraction expansion, we can write the transfer function as

$$H(z) = \frac{8z}{z-2} \frac{10z}{z-3} \frac{3z}{z-1}$$

The block diagram is shown in Figure 6.16. From Figure 6.16, we can see that the states are

$$v_1(n+1) = x(n) + 2v_1(n)$$

$$v_2(n+1) = x(n) + 3v_2(n)$$

$$v_3(n+1) = x(n) + v_3(n)$$

and the output is

$$y(n) = 8x(n) + 16v_1(n) + 10x(n) + 30v_2(n) + 3x(n) + 3v_3(n)$$

The states and output equations in matrix form are

$$\mathbf{v}(n+1) = \begin{pmatrix} 2 & 0 & 0 \\ 0 & 3 & 0 \\ 0 & 0 & 1 \end{pmatrix} \mathbf{v}(n) + \begin{pmatrix} 1 \\ 1 \\ 1 \end{pmatrix} x(n)$$

$$\mathbf{y}(n) = \begin{pmatrix} 16 & 30 & 3 \end{pmatrix} \mathbf{v}(n) + (21)\mathbf{x}(n)$$

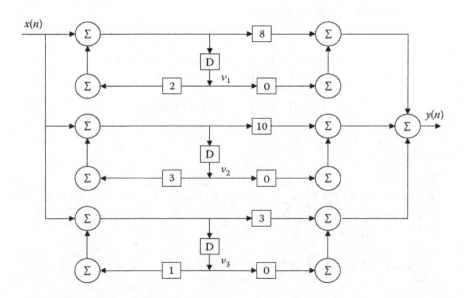

FIGURE 6.16 Block diagram for EOCE 6.11.

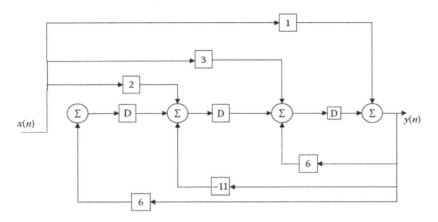

FIGURE 6.17 Block diagram for EOCE 6.11.

4. We can represent the system in the block diagram shown in Figure 6.17. Let the output of the first delay be $v_1(n)$. Then

$$v_1(n+1) = 6y(n) = 6[x(n) + v_3(n)] = 6v_3(n) + 6x(n)$$

Let the output of the second delay be $v_2(n)$ and the output of the third delay be $v_3(n)$. Thus,

$$v_2(n+1) = 2x(n) + v_1(n) - 11v_3(n) - 11x(n)$$

$$v_3(n+1) = 3x(n) + v_2(n) - 6v_3(n) + 6x(n)$$

The output from the figure is given by

$$y(n) = x(n) + v_3(n)$$

The state and output equations are then

$$\mathbf{v}(n+1) = \begin{pmatrix} 0 & 0 & 6 \\ 1 & 0 & -11 \\ 0 & 1 & 6 \end{pmatrix} \mathbf{v}(n) + \begin{pmatrix} 6 \\ -9 \\ 9 \end{pmatrix} \mathbf{x}(n)$$

$$\mathbf{y}(n) = \begin{pmatrix} 0 & 0 & 1 \end{pmatrix} \mathbf{v}(n) + (1)\mathbf{x}(n)$$

We will use MATLAB next as in EOCE6_11 to find the step response, $y(n)$, for all four representations and show that $y(n)$ for all four representations is the same. Let us also use zero initial conditions.

The plots are shown in Figure 6.18. Note in this example that the **A** matrix in the third case is diagonal. So for analytical solutions it is desirable that you do partial fraction expansion and then draw the block

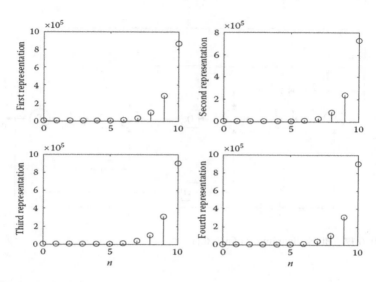

FIGURE 6.18   Plots for EOCE 6.11.

diagram from which you will obtain the **A** matrix in its diagonal form
Note also that if

$$A = \begin{pmatrix} a_1 & 0 & 0 \\ 0 & a_2 & 0 \\ 0 & 0 & a_3 \end{pmatrix}$$

then

$$A^n = \begin{pmatrix} (a_1)^n & 0 & 0 \\ 0 & (a_2)^n & 0 \\ 0 & 0 & (a_3)^n \end{pmatrix}$$

6.10   END-OF-CHAPTER PROBLEMS

EOCP 6.1

Let

$$A = \begin{pmatrix} 0 & 0 & 0 \\ 0 & 0 & 1 \\ -3 & -2 & -6 \end{pmatrix} \text{ and } B = \begin{pmatrix} 0 & 0 & -1 \\ 1 & 0 & -2 \\ 0 & 1 & -3 \end{pmatrix}$$

1. Find A^2, A^3, A^{-1} and the eigenvalues and eigenvectors for **A**.
2. Find B^2, B^3, B^{-1}, eigenvalues and eigenvectors for **B**.

EOCP 6.2

For

$$B = \begin{pmatrix} -2 & -3 & -4 \\ 0 & 0 & 1 \\ 0 & 1 & 0 \end{pmatrix}$$

1. Find eigenvalues and eigenvectors for B.
2. Form the matrix P, which has the eigenvectors as its columns.
3. Find $P^{-1}BP$. Is it diagonal?
4. Find the eigenvalues for $P^{-1}BP$.

EOCP 6.3

With

$$A = \begin{pmatrix} 2 & 0 & 0 \\ 0 & 3 & 0 \\ 0 & 0 & 4 \end{pmatrix} \quad B = \begin{pmatrix} 0 & 0 & 0 \\ 0 & 0 & 1 \\ -2 & -1 & -3 \end{pmatrix} \quad C = \begin{pmatrix} 0 & 0 & -1 \\ 0 & 1 & -2 \\ 0 & 0 & -3 \end{pmatrix}$$

Find A^n, B^n, and C^n.

EOCP 6.4

Consider the following difference equations:

1. $y(n) + a_1 y(n-1) + a_2 y(n-2) = b_0 x(n)$
2. $y(n) + a_2 y(n-2) = b_0 x(n)$
3. $y(n) + a_3 y(n-3) + a_4 y(n-4) = b_1 x(n-1)$
4. $y(n) + a_1 y(n-1) + a_2 y(n-2) = b_0 x_1(n) + b_1 x_2(n)$
5. $y(n) - a_4 y(n-4) = b_0 x_0(n) + b_1 x_2(n)$
6. $y(n) - y(n-3) = x(n)$

Find the state-space representation for each system given earlier. Find A, B, C, and D.

EOCP 6.5

Given the following impulse responses

1. $h(n) = (.1)^n u(n) + (-.1)^n u(n)$
2. $h(n) = (.1)^n \sin(n) u(n)$
3. $h(n) = n(.1)^n \cos\left(\frac{3\pi}{2} n\right) u(n)$
4. $h(n) = (n-1)(.1)^{n-1} u(n-1)$
5. $h(n) = (.1)^n u(n) * (.5)^n u(n)$

Find the state-space representation for each system. Find A, B, C, and D.

EOCP 6.6

Consider the following blocks in Figures 6.19–6.23. Find the state-space representation for all blocks. Find **A**, **B**, **C**, and **D**.

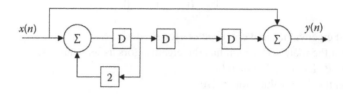

FIGURE 6.19 Block diagram for EOCP 6.6.

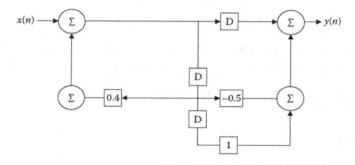

FIGURE 6.20 Block diagram for EOCP 6.6.

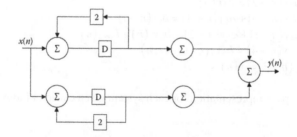

FIGURE 6.21 Block diagram for EOCP 6.6.

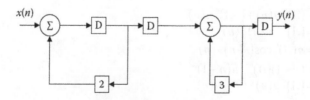

FIGURE 6.22 Block diagram for EOCP 6.6.

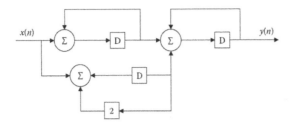

FIGURE 6.23 Block diagram for EOCP 6.6.

EOCP 6.7

Consider the following transfer functions:

1. $H(z) = \dfrac{(z-1)(z-.5)}{(z+1)(z+.5)}$

2. $H(z) = \dfrac{z}{(z+1)(z+.5)(z-1)}$

3. $H(z) = \dfrac{(z-1)(z-.5)}{(z+1)}$

4. $H(z) = \dfrac{z^2 - 1}{z^3}$

5. $H(z) = \dfrac{z^3 + 3z^2 + 2z + 1}{z^3 + 4z^2 + 2z + 3}$

Find four different state-space representations for each transfer function.

EOCP 6.8

Consider the system

$$y(n) + ky(n-1) + 3y(n-2) = x(n)$$

Use the state space in all parts.

1. For what value k is the system stable?
2. Take a value for k that makes the system stable and find the eigenvalues of the system.
3. For the value of k in part 2, find the output $y(n)$ for $x(n) = u(n)$ using MATLAB.
4. Repeat part 3 in the list using the z-transform method.
5. Find the transition matrix A^n.
6. Find $H(z)$ from the state equations analytically.
7. Find $H(z)$ using MATLAB.
8. Use MATLAB to find $h(n)$ from $H(z)$. Find the residues using MATLAB.

EOCP 6.9

Consider the system

$$y(n) + 0.1y(n-1) + y(n-2) = x_1(n) + x_2(n)$$

with $y(-1) = 0$ and $y(-2) = 1$.

1. Is the system stable?
2. Put the system in state space.
3. Use MATLAB to find $\mathbf{h}(n)$ using the state equations.
4. Use MATLAB to find $y(n)$, the step response.
5. Use MATLAB to find the derived initial conditions with $x_1(0) = x_2(0) = 0$.
6. Find $\mathbf{H}(z)$ using MATLAB and identify each entry with the $\mathbf{H}(z)$ matrix.
7. Find $\mathbf{A}^n$.

EOCP 6.10

Consider the following systems:

$$H(z) = \frac{z+2}{z^2 + z + 1}$$

$$H(z) = \frac{1}{z^2 - .6z + .05}$$

For both systems using the state-space method

1. Find two state-space representations.
2. Find $h(n)$.
3. Find $\mathbf{A}^n$.
4. Find the step response.
5. Check stability for both systems.

EOCP 6.11

Consider the following system:

$$H(z) = \frac{z^2 + z + 1}{z^2 + kz + .05}$$

1. Write the state-space representation.
2. For what value(s) of k is the system stable?
3. Find the difference equation representing $H(z)$.
4. Pick a k that makes the system stable and find the output $y(n)$ for $x(n) = n$ $\sin(n)u(n)$. Use the MATLAB function dlsim to do that.
5. Find $\mathbf{A}^n$ for a given k.

EOCP 6.12

Consider the system

$$y_1(n) - y_2(n-1) = x_1(n)$$

$$y_2(n) - y_1(n-1) = x_2(n)$$

where $y_1(n)$ and $y_2(n)$ are the outputs and $x_1(n)$ and $x_2(n)$ are the inputs.

1. Write the state-space equations describing the system.
2. Is the system stable? Find the eigenvalues for **A**.
3. With $x_1(n) = x_2(n) = u(n)$, find $y_1(n)$ and $y_2(n)$.
4. Find the transfer function matrix $H(z)$ from **A**, **B**, **C**, and **D**.
5. Find the state transition matrix.
6. Find the impulse response of the system.
7. Draw the block diagram for the system and obtain a different state-space representation.
8. Find the step and impulse responses for the representation in part 7 in the list.
9. What are the eigenvalues for the new representation in part 7 in the list?

EOCP 6.13

Consider the system in Figure 6.24.

1. Write the state equations for this system.
2. Is the system stable?
3. What is the transition matrix?
4. What is the transfer matrix $H(z)$?
5. Find $y_1(n)$ and $y_2(n)$ for $x(n) = 10 \cos((2\pi n/3) + .1)u(n)$.
6. Find the impulse response for the system.
7. Write the coupled two difference equations describing the system.

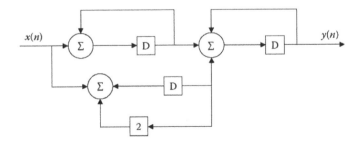

FIGURE 6.24   System for EOCP 6.13.

EOCP 6.14

Consider the following systems:

$$(a)\ H(z)=\begin{pmatrix} \dfrac{z}{z^2+z+1} & \dfrac{z^2+1}{z^2+z+1} \\[3mm] \dfrac{z+1}{z^2+z+1} & \dfrac{z^2+z}{z^2+z+1} \end{pmatrix} \quad (b)\ H(z)=\begin{pmatrix} \dfrac{z}{z^2+1} & \dfrac{z+1}{z^2+1} \end{pmatrix}$$

1. Draw the block diagram for these systems.
2. What are the state-space representations for both systems?
3. What is the transition matrix for both systems?
4. Find the step response for both systems.
5. Find the impulse response for both systems.
6. Are the systems stable?

Index

Note: Page numbers in *italics* indicate figures and **bold** indicate tables in the text.